Life in the Universe
and Where to find it

Life in the Universe, and Where to find it

Copyright © 2011 by Kevin Shoemaker

This book is printed on acid free paper

A Shoemaker Labs Book
Lafayette, Colorado

e-mail: Shoemakerlabs@gmail.com

ISBN 978-0-9815092-5-9
ISBN 0-9815092-5-8

Registered with the Library of Congress

Cover Art courtesy of Atlasoftheuniverse.com
Chapter Artwork courtesy of NASA, the Shoemaker family and Wikipedia

First edition October, 2011
Printed in the United States of America

To Judi, Leah and Stephen
for their inspiration and support

Acknowledgment

I would like to sincerely thank Jason Coder for his editing and comments. Also, I would like to thank my friends Gary, Eliot, and Russ for their editing and comments. Finally, I would like to thank my lovely wife Judi, daughter Leah and son Stephen, for their encouragement and patience.

Author's note

I started early wondering about life in outer space. By high school I had a Jupiter antenna array listening to the noise bursts coming from the planet with a low quality receiver that amazingly produced very clear recordings of this extraordinary amount of energy that Jupiter was beaming towards Earth. I studied philosophy and astronomy in college with the hopes of turning into a Cosmologist. The many books on the subject convinced me that this was the ultimate field of study, answering probably the most important questions that needed to be asked like 'Are we alone?'

But something was bugging me about the studies that had been undertaken to find life, both then and now. One problem is that the distances between stars is very great. Another problem that concerned me was we were using instruments that by simple calculation appeared to be far inadequate for the mission. Finally, I was not convinced that the presupposition that other life would be broadcasting in similar enough ways to our own transmitters for our receivers and antennas to detect. This last issue seemed to border on ethno-centricity.

And what really is life? What percentage is sentient? Will life in the Universe be similar to that on Earth?

From a scientific perspective, we have a model that works. Earth, where it is, with the appropriate chemical compositions, harbors life with ease, and has done so for billions of years.

From a philosophic perspective, most of us feel intuitively that life elsewhere in space will be found. We project these feelings into science fiction literature and movies. We feel so strongly about this sometimes that the question of extraterrestrial life is sometimes answered by asking questions like "is your car parked outside where you left it 10 minutes ago?" "Of course" we say to others, although we might not have any physical proof of it (this assumes we cannot see the car at the time). In the same fashion we ask "Is there life out there?" "Of course we answer, how could there not be?" Is this a valid approach to answering this very important question of sentient extraterrestrial life? Maybe....or maybe not.

It appears that we are now on the precipice of answering this question in a more substantive fashion. As a result of recent research and discoveries, I predict that the following sequence of events will take place based on current knowledge, both scientific and philosophical.

Planets will be found in abundance

Many of these planets will exhibit chemical signatures in their atmospheres that could be generated by living organisms

Further careful and long term studies of many of selected candidate planets will show enough indicators and chemical life cycle changes that the astronomers will establish high confidence levels as to the probability of living organisms

Finally, someone will say "prove it otherwise" and when this task ends in futility we will then be convinced

But the new life will not be like our own

This book will take you through the steps, both scientific and philosophic towards a better understanding of who we are and where there will be life in the Universe. I hope you enjoy reading it as much as I enjoyed writing it.

K. Shoemaker July, 2011

Table of Contents

Chapter 1 - Introduction

Time is getting close for the detection and confirmation of other life besides our own in the Universe, but we need to understand the basics of accomplishing this first. These basics include a rigorous examination of how much power is required for a signal to reach the Earth from candidate planets and be detected with reasonable instruments. This is a multi-faceted problem that will be examined in this book. The details include frequency selection, antenna selection, opportunistic pointing of the radiation towards us and opportunistic pointing of our instruments towards the emitters. Quickly the reader will realize that even with our "loudest" emanations, a close by world would have to be extremely well prepared as well as extremely lucky to hear us.

We are overly influenced by Science Fiction and Fantasy and as a result, expect life out there to basically look like us and speak our language. This will not be the case. The fact is, life out there will be mostly non-sentient and if any life has evolved to the point of consciousness, the probability of having it in the form of homo sapiens is extremely low. This is because the influence of the environment and history of a candidate planet will never be a copy of our own. Chemistry, meteor impact history and living conditions on that planet cannot ever be identical to ours either. One cannot assume that each candidate planet will have for instance, oceans covering 4/5 of the globe or our exact temperature range or gravity constant.

But the exercise of looking out there for a sign, looking out there for the first confirmation of biology does teach us something. And that is that higher intelligence includes (or maybe even mandates) curiosity. Assuming this is universal, higher intelligence out there should be curious about other life as well. This will be a helpful attribute once we make it there (or they come here). Due to the laws of physics, especially the speed of light, the most reasonable way of traveling to the stars will be with a very large ship containing people committed to spending at least a generation in space. Realistically, even at speeds close to that of light, many years will be required to visit even the closest stellar neighbors. We need to choose carefully when we make our first voyages. Also, there is the high chance that some of us will colonize other planets, then attempt to communicate over the vast distances.

And how would we communicate back to Earth? Better yet, how would an intelligent life form communicate? One needs to know the basics of electromagnetic propagation first. Even at the very short distance of 10 light years away, the task will be daunting. They will have to be able to generate substantial amounts of power and use a very large radiating aperture. Then they will have to wait at least 20 years for a reply.

Cogent approaches must be considered now, not just radio or optical communications. The electromagnetic spectrum harbors important information across its entire expanse. It has been only in the last few decades at emission lines in the radio, microwave, millimeter wave, sub-millimeter wave, infrared, optical and ultraviolet have been discovered and appreciated. All chemicals and elements have detectable resonances in the spectrum. Thousands of these "lines" have been observed by radio and other types of telescopes. Many thousands chemical compositions have been examined in a laboratory setting and shown to contain these lines. This detail will be an important contributor to the ultimate discovery of life in the Universe other than our own. The following pages will elucidate these discoveries and some of the amazing results from the science. These include mapping the Universe in a particular chemical, which has been done for many chemical species. Also, because these chemicals are subject to the Doppler effect, a researcher can detect the motions of the associated clouds of particular gases. In this way radio astronomers were able to map out the Milky Way as if from looking above it, even though we have a limited view based on our position in one of its arms.

The current planet location technologies (Doppler and Photometry) will combine with spectroscopic techniques to increase the likelihood of the detection of life. Specifically, planets are being discovered now by two main techniques. The first is a survey style examination that compares candidate star systems over a long period of time to evaluate whether or not the line emission spectra from the star is shifting. By "graphing" the line shifts over time, an astronomer can determine any periodicities that might be present. As the candidate star moves towards and away from us (we subtract any of our own movements) due to planets surrounding it, one can determine the number and orbital period of these planets. Another approach is to examine and record the light level of a candidate star and observe any dips in the readings. In this way, planets that move in front of the star and occult the light are revealed. Again the length and depth of the occultation indicated the period and size of the planet. Multiple dips with different periods and light curve depths indicate multiple planets.

Life is simple initially and to assume that there will be many like us is folly. This concept will be explained in detail further into this book. As the reader will discover, our planet is unique in terms of the history of its life. The chemical makeup is unique, the interaction with various forms of space debris over time is unique, the evolution of the plants and animals has been unique. It stands to reason that all other extra solar planets that have been discovered and that will be discovered will follow different paths than our own. There might however be some gross similarities. These would include the supposition that life is simple initially. In later chapters, the Miller - Urey experiments in 1952 will show that simple life precursors in the form of amino acids can be manufactured in the laboratory by simulating the primordial Earth with its lightning bolts, volcanic atmosphere and water. The theory of abiogenesis and panspermia will be explained as it relates to the evolution of more and more complex life forms culminating in sentient life.

Even now the vast majority of life is simple. Most is single cell, followed by crude conglomerations in the form of algae, lichen and mosses. As can be seen within the time lines in later chapters, the more complex forms occurred in the most recent past. It also is interesting that the most complex forms of life, where intelligence appears and becomes sentient, occurred in the last .1% of the timeline of life. Its hard to imagine where life will be in just another million years.

Assuming that rate of evolution is the same, then distance to habitable planets must be taken into account (e.i. # of light years away). Many of our stellar neighbors were created in about the same time. This is a consistent theme in stellar evolution where star nurseries produce a large amount of offspring, roughly at the same time. Another consistent theme in astronomy is that new stars usually have a planetary disk. It stands to reason as the new star is spinning and therefore creating a disk of material which coalesces into individual planets. Whats left is a plethora of smaller chunks of materials like asteroids and comets which are captured by the planets's gravities, pelt the surface or burn up in any atmospheres present. After the "cosmic cleanup" is mostly complete, life can take root.

If we allow ourselves to believe that our stellar neighbors, being mostly the same age, have planets in a habitable zone and have roughly the same chemical makeup, we can then feel comfortable in the fact that life will evolve on these planets and move from simple to complex, just like on Earth. What we cannot do however, is assume that all planets will follow the same paths, there are just too many variables to consider. For instance, I will mention just one...what if the candidate planet has an ocean a fraction the size of our own. Maybe just the extent of the Great Lakes in North America. Would life on that planet mirror our own? Of course not. Too many of Earth's lifeforms came from our oceans and there is ample evidence that the first land crawlers emerged from these seas.

What's now left is that if life within our stellar neighborhood is evolving at roughly the same pace as ours then the distance between other solar systems and ours becomes important. In other words, if there really is a chance that there is sentient life on a planet at a distance of say 100 light years, and they happen to understand electromagnetic theory and happen to have designed and built large radio telescopes and have also been lucky enough to point these telescopes towards Earth and then lucky enough to interpret our emanations....then at best they will be able to send a signal and have us receive it (assuming we are looking at the exact right place when they do) within 200 years. Why? Because we have only been transmitting for a little over 100 years, and at a low level by the way, and if they respond immediately, then we will hear them repeat the Titanic's SOS a hundred years from now. There are a lot of "ifs" in our assumptions and therefore its not surprising that we have not heard anything in terms of intelligent

9

communications sent our way. Remember Fermi's paradox...."Where are they?"

Fermi posed his question during an informal discussion in 1950. Most people then and most now feel that if there are truly billions upon billions of stars, there has to be intelligent life on some of their planets. Even if a small percentage is sentient, that is still quite a lot of planets presumably capable of communication. Fermi asked, 'so if there are so many planets harboring intelligent life, where are they? Why have they not communicated with us?' There is an answer to this paradox. It comes in two parts; they are too far away and we are mistaken to assume that intelligent life will be just like our own.

In the case of the first part, it will be shown clearly in this book that the signal strength loss for even the closest stars requires extraordinary efforts to overcome. The larger the signal loss, the bigger the antenna required to receive the signal and the bigger the magnitude of power required to overcome the distance. The reality is that the bigger the antenna, the smaller the beamwidth. In other words just like increasing magnification in a microscope, whereby smaller and smaller details can be examined, so too the effect of increasing the size of an antenna or telescope. Smaller and smaller details are due to smaller and smaller beamwidths or fields of view. As a consequence, many more fields of view have to be monitored to have a chance to catch a signal from a distant source. This takes more and more time to cover an area of sky or we spend less and less time on a particular part of the sky as we move from field of view to field of view. Is there an alternative? Yes, and it be discussed in later chapters.

Life in the Universe and Where to find it

The second issue is that humans are under the mistaken impression that when we discover life out there, especially sentient life, it will be just like going to a foreign country. This is far from the truth, as life can take on an almost infinite number of forms. Sentient life could be in plants, or fish, or birds or....you get the idea. Due to the infinite number of environments out there, and how life adapts to these environments, any similarity between our set of species and theirs will be extremely remote. It is important to realize that our model, life on Earth, works and is highly successful, showing how easy it is to create millions of species of plants and animals. These life forms have been appearing and evolving for billions of year. It is only during the last very small percentage of life's history that consciousness has been evident. This begs the question of what we will be in another few million years, or even more excitingly, what cultures a few million years ahead of us in the Universe are like now. Predictably, they will not be like our own. We will go into details in later chapters here as well.

The distance between planets causes a problem, that is, that there is a signal loss due to distance that follows the rule of $1/R^2$. This is true for any electromagnetic wave, which includes radio, optical, infrared, ultraviolet, etc. This signal loss is better referred to as a component of a link budget. This we will find includes the power of the transmitting signal, the gain of the antennas in the link, the receiver sensitivity and the bandwidth of the signals as well as the signal loss due to distance. We will find out a few amazing things, first that it is very hard and optimistic to create a communications link between stars, even close ones. The second amazing thing is that natural phenomena, like Jupiter's decametric emissions, have the necessary power and "antenna" gain to perform the feat. We will also discover that life has a footprint that can be detected by radio telescopes, not in the form of radio or TV broadcasts but by the chemical changes evident in the atmospheres of planets. These chemical properties come in the form of radiation and absorption, both detectable by telescopes (all types).

We are moving towards answers now, in a cogent way but we have a lot of work ahead of us.

Can we predict where and when this discovery will be made? Possibly. We certainly can bracket in the required technologies and expected time lines. We can see together what will be required and realize now, that the 'end of the tunnel' is in sight. To for instance, make a prediction on the exact date will be impossible, but to say that it will be within a generation or so will be quite reasonable.

We will discuss both light and radio detection possibilities. We will also discuss several forms of electromagnetic radiation and their usefulness for communication purposes. We will show that listening to broadcasts on other planets is very similar to listening to people talk in New York City, from San Francisco. A good and optimistic scientist or engineer will say that under the right circumstances, this can be done, but the details will be overwhelming and impractical.

The philosophy of extrasolar life searches needs to be discussed as well. As the origins of these speculations are from the ancient Greek philosophers. Although humans might have wondered about the possibilities before Pre-Socratic Philosophers, it was they that formalized the work and documented their thoughts.

Later philosophers like Kant discussed many details about the probabilities of life out there and felt that it was in intuitional truth, like how we know that the sun will come up tomorrow, that gave us the mental strength to know life existed elsewhere in the cosmos. We will find this to be presumptuous. Current philosophic thought on this subject include the question: "Who cares?" We must answer this one as well.

This book is written for those readers with a reasonable amount of knowledge in science and an interest in philosophy. Although there are a few equations, the math is not hard to follow and is present to give legitimacy to the conclusions. The coverage of the philosophic aspects is admittedly cursory and further interest by the reader will be rewarded with many more reflections on this subject. It has been said that Science Fiction in general and Space Exploration in particular are some of the last bastions for philosophers, as they are the ones that can guide us into understanding what we discover.

Chapter 2 - History of Speculation

Hipparchus

"No great discovery was ever made without a bold guess"
-Newton

Time Line

Although there is significant evidence of civilizations going back to 12,000 years ago using stellar objects to predict seasons and harvest time (and thus discover the rudiments of astronomy) it was not until Thales of Miletus in pre-Socratic Greece that open speculation about other life in the Universe began.

Thales (624 – 546 B.C.)

In approximately 600 B.C. the Greek philosopher Thales experimented with amber and silk to create sparks and attract small bits of organic materials like straw and feathers. He found that rubbing the silk and amber would create varying amounts of attraction or sparks. He also found that loadstone had varying degrees of magnetic attraction. His scientific observations of amber and loadstone allowed him to understand the very basics of electromagnetic theory as it related to electricity and magnetism. These two entities are intertwined to allow radio waves to be created and propagate. The Greek word for amber is *elektron* and the word for magnetism is *magnesia*. From his observations and writings we now use terms like magnets and electronics.

He took careful notes on physical phenomena including meteorological and astronomical events. Thales also made fundamental observations in mathematics and in fact, he was the first (or one of the first) people to predict a solar eclipse. He spent time in Egypt to learn geometry and surveying, then examined the Babylonian records of the motion of the stars and sun. His notes revealed a periodicity in the solar and lunar eclipses, allowing him to be the first to predict an event.

Thales is known as the first philosopher, he theorized that all things are derived from water. Because this belief had a natural and not a supernatural origin, it led other philosophers to take this rationalist thought process and use it to reexamine their understanding of the essence of matter and the nature of objects in the sky. From this new philosophy, ideas about extraterrestrial life came into being. He was a student of the Delphic Oracles, women who answered questions about life and the origin of the universe.

In addition, he understood weather patterns enough to predict a particularly good olive harvest one year. He purchased most if not all of the olive presses in the local region of Greece and apparently made a small fortune when the olives needed to be turned into olive oil.

As a philosopher, he believed that philosophy was the trunk of knowledge with the branches being subjects like mathematics, astronomy and biology. From these branches would grow more branches as major scientific subjects would be broken down into sub-categories. For instance, biology is now broken down into categories like microbiology and ecology. Philosophers who followed Thales, like Plato and Aristotle, furthered the notion of the tree of knowledge and contributed much to describing scientific phenomena and characterizing intellectual thought.

The important transformation ushered in by Thales was that he suggested that the stars were not "chariots of the Gods," which was the prevailing theory. Alternatively, he thought that they were possibly suns like our own. Thus, he separated the supernatural from the natural and allowed more scientific approaches to be applied to the cosmos.

Thales of Melitus

Other Philosophers and Astronomers

After Thales came his student Anaximander and many others who defined the Universe in more and more scientific details. The motions of the planets were examined and thought about in detail. The Greeks referred to them as 'wanderers'.

Hipparchus of Nicaea (190 – 120 BC) is widely accepted as the first astronomer and the creator of trigonometry, he designed astronomical instruments and made careful measurements of the positions of the stars and motions of the planets. He also came up with a technique to measure the brightness of stars that is surprisingly close the the present log based methods used today. There have been orbiting observatories named after him as well as a primary star catalog.

Aristotle (384 – 322 B.C.) , student of Plato (424 – 348 B.C.) formulated a theory of the Universe that became the standard until the 16th Century. This theory, geocentricism postulated that the Earth was the center of the Universe. During this extended period, speculation about extraterrestrial life was more religion based than scientific. I wasn't until Copernicus and later Galileo, whose observations challenged the geocentric idea, that new ideas about the cosmos started to emerge again.

Aristotle denied that there could be a plurality of worlds and seemingly rendered extraterrestrial life philosophically untenable.

Aristotle

Ancient and medieval ideas

Generally, during the period of the Greeks and from that point on, several themes developed about the Universe. These themes were based on religious as well as secular beliefs. I was common during this early period to believe that there were many worlds populated by intelligent, non-human life-forms, except that these worlds were mythological and not part of the heliocentric or sun centered construction of the solar system. In this scenario, the sun was not considered or understood to be one of many stars in the Universe. For instance:

19

Life in the Universe and Where to find it

There were fourteen *loka* in Hindu beliefs. They also believe in endless repeated cycles of life that have developed multiple worlds of existence and their mutual contacts. There are also many universes where souls can contemplate the purpose of life

Nine worlds of Old Norse mythology

The atomists of Greece like Epicurus argued that an infinite universe should have an infinity of populated worlds.

The Jewish *Talmud* states that there are 18,000 other worlds, whether or not these are physical or spiritual worlds is not indicated. Following this is an 18th century writing *Sefer HaB'rit* that suggests that extraterrestrial creatures exist, and may possess intelligence. It also states that humans should not expect these creatures from other worlds to resemble life on Earth.

According to Islamic beliefs, the *Quran* has proof of life on other planets and that at some future time, these entities will meet with those on Earth.

In Christianity, the Ptolemaic system was widely accepted and the idea of alien life is, or was, considered aberrant. In 1277 Bishop Eienne Tempier overturned the
Aristotelian/Ptolemaic system by saying that God could have have created more than one world, given his omnipotence. Further, he argued that aliens actually existed but were rare. Cardinal Nicholas of Kues speculated about aliens on the Moon and Sun.

In these multi world scenarios, the Sun and Moon after were considered vehicles that were driven by gods. An example would be in the Japanese fold tale *The Tale of the Bamboo Cutter*, written in the 10th century, where the princess of the Moon people visiting Earth.

Early modern period

Once the telescope was invented and the ideas of Copernicus were espoused, the Ptolemaic or geocentric cosmology was becoming passé. Soon, it was clear that the Earth was but one planet amongst countless bodies in the Universe. The idea of extraterrestrial life could now move into the scientific mainstream.

Copernicus

Life in the Universe and Where to find it

Notable philosophers and writers:

Giordano Bruno, who in the 16th Century argued for an infinite Universe in which every star is surrounded by its own planetary system. He wrote that other worlds "have no less virtue nor a nature different to that of our earth and contain animals and inhabitants."

Anton Maria Schyrleus of Rheita, a Czech astronomer in the 17th century thought "if Jupiter has inhabitants, they must be larger and more beautiful than the inhabitants of the Earth, in proportion to the [characteristics] of the the two spheres".

Cyrano de Bergerac in his book *The Other World: The Societies and Governments of the Moon* where he considered extraterrestrial societies were "humoristic" or ironic parodies of earthly society.

Henry More, who wrote the essay "Democritus Platonissans, or an Essay Upon the Infinity of Worlds" in 1647.

Sir Richard Blackmore wrote "The Creation: a Philosophical Poem in Seven Books" in 1712 and observed "We may pronounce each orb sustains a race of living things adapted to the place. He also suggested "Our world's sun becomes a star elsewhere."

Fontanelle wrote "Conversations on the Plurality of Worlds" translated into English in 1686. He discussed the possibility of extraterrestrial life, expanding, rather than denying , the creative sphere of a Maker.

Benjamin Franklin as well speculated about the existence of life in the cosmos.

Garrett P. Serviss wrote *Edison's Conquest of Mars* in 1897.

William Whewell of Trinity College, Cambridge popularized the word *scientist* and theorized that Mars had seas, land a possible life forms.

Bruno *De l'Infinito Universo et Mondi, 1584*

Immanuel Kant (1724-1804) in his book 'General History of Nature and Theory of the Heavens' discussed his "Nebular Hypotheses" where he deduced that the Solar System was formed from a cloud of gas or nebulae. He also deduced that the Milky Way was a large disk of stars that was formed from a much larger spinning cloud of gas. Finally, he speculated that there would be many similarly large and distant disks of stars in the Universe. These speculations, amazing at the time of little astronomical instrumentation, proved to be true.

Kant is immensely important in the history of philosophy and continues today to be relevant. His most important book *Critique of Pure Reason* united reason with experience to end the era of speculation. He said "it always remains a scandal of philosophy and universal human reason that the existence of things outside us should have to be assumed merely on faith, and that if it occurs to anyone to doubt it, we should be unable to answer him with a satisfactory proof." Also he said "Up to now it has been assumed that all our cognition must conform to the objects; but ...let us once try whether we do not get farther with the problems of metaphysics by assuming that the objects must conform to our cognition."

These were revolutionary ideas that set the foundation for scientific processes and methodologies.

Kant

A contemporary philosopher, **Francis Seeburger** (1946 -), once asked the question, "Who Cares?" when queried about the existence of life in the Universe. To a scientist or other technical person, this question would appear silly. However it is not, as we must look in many places to answer it properly. The first place is certainly within ourselves as human beings, where our drive to discover and need to understand forms the basis for an answer to why we care. As a society both locally and globally, it seems important as well. We could of course have never cared, but we always did, and maybe this is why we continue to discuss, plan and experiment on the subject. It is certainly a truism that the subject will never die, and that the quest to find out if we are alone or surrounded with life will continue until resolved. Dr. Seeburger specializes in Phenomenology and in particular Martin Heidegger. As a result he is an expert on modern philosophic thought. His is also an author of several critically acclaimed books on the subject.

Seeburger

Life in the Universe and Where to find it

Cosmologists

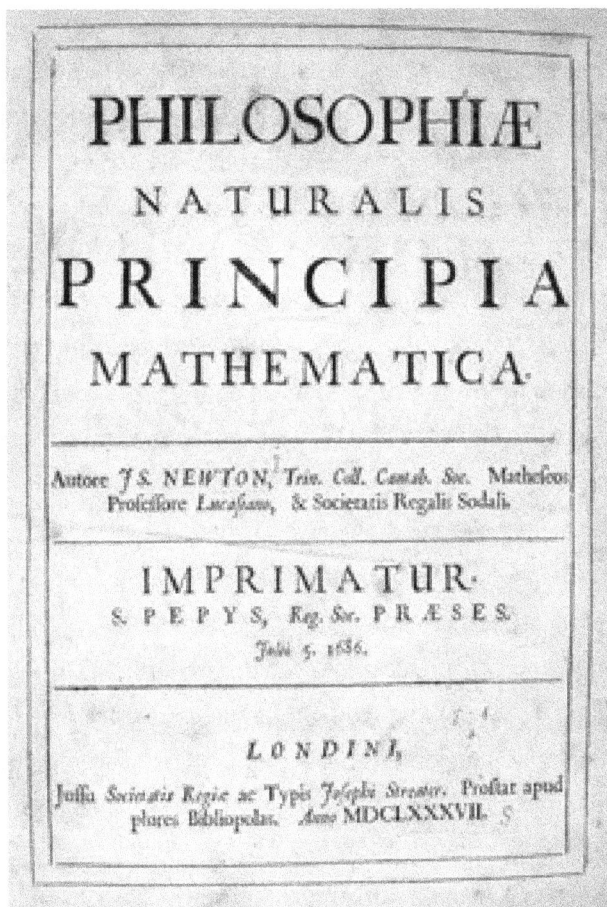

PHILOSOPHIÆ
NATURALIS
PRINCIPIA
MATHEMATICA.

Autore JS. NEWTON, Trin. Coll. Cantab. Soc. Mathefeos
Profeffore Lucafiano, & Societatis Regalis Sodali.

IMPRIMATUR·
S. PEPYS, Reg. Soc. PRÆSES.
Julii 5. 1686.

LONDINI,
Juffu Societatis Regiae ac Typis Josephi Streater. Proftat apud
plures Bibliopolas. Anno MDCLXXXVII.

I was like a boy playing on the sea-shore, and diverting myself now and then
finding a smoother pebble or a prettier shell than ordinary, whilst the great
ocean of truth lay all undiscovered before me

-Issac Newton

26

Many cosmologists or scientists writing about cosmology have written good books and papers that address the problems of interstellar communications and the probable existance of life in the Universe. What follows is a small sampling of some of these researchers and their main interests. Many of the ideas shown here are still valid and many are becoming more valid as more information and research become available.

Isaac Newton (1642 – 1727) was a natural philosopher who wrote the book *Philosophiae Naturalis Prinicpia Mathematica* published in 1687, which lays the foundations for most of our knowledge on classical mechanics and universal gravitation. In this work, Newton described the concept of gravity and the three laws of motion which dominated the scientific view of the physical universe for the next three centuries. Newton showed that the motions of objects on Earth and of celestial bodies are governed by the same set of natural laws, by demonstrating the consistency between Kepler's laws of planetary motion and his theory of gravitation, thus removing the last doubts about heliocentrism, which was the Sun centered solar system theory The *Principia* is generally considered to be one of the most important scientific books ever written.

Newton also created the mathematical form called Calculus and wrote a classic textbook on optics.

Newton

William Herschel (1738 – 1822) was a German born astronomer who became famous for building very large (for that day) telescopes and discovering Uranus as well as its major moon, Titania and Oberon. In addition, Herschel discovered two moons of Saturn and is credited with discovering infrared radiation.

To Herschel, life was probably on Mars, as well as the rest of the planets in our solar system.

Herschel

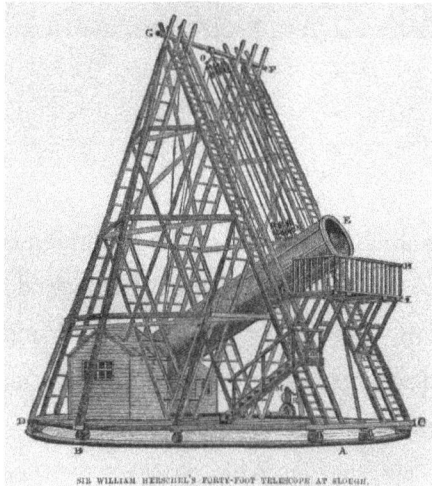

Herschel's Telescope

Life in the Universe and Where to find it

Percival Lowell (1855 – 1916) was the son of a wealthy Boston businessman who moved to Flagstaff, Arizona to establish a world class observatory. Now known as the Lowell observatory, it is one of the centers of planetary astronomy in the world. Lowell observed Mars primarily, wrote several books about the subject and considered the canals on the planet as proof of life. He thought these canals were an artifact of an ancient civilization making a desperate last effort to survive. This idea influenced H.G. Wells' book *The War of the Worlds.* Wells expanded on this idea and envisioned creatures from this dying planet invading Earth.

14 years after his death, another astronomer working at Lowell Observatory, Clyde Tombaugh, discovered Pluto. The choice of the name and its symbol (PL) were influenced by Lowell's initials.

Spectroscopic analysis of Mars' atmosphere began in 1894 and proved conclusively that water and oxygen were not abundant enough to support the idea of canals. By 1909, a fortuitous perihelic opposition of Mars put an end to the canal theory as the planet was at its closest point to Earth and thus close enough for a detailed examination. Most of this work was done by William Wallace Campbell, an America astronomer.

Life in the Universe and Where to find it

Lowell

Albert Einstein (1879 – 1955) revolutionized our understanding of the Universe. Advancing beyond Newton's equations, he formulated new concepts like Special Relativity, General Relativity, and the Unified Field Theory that apply to the actions of our Universe. To this day, his equations offer insights into Black Holes, Gravitational Lensing, Quantum Mechanics and many other physical phenomena. There are many more discoveries waiting in his equations as modern scientists use these as the basis for explaining phenomena from the atomic to the astrophysical. Although he did not spend a significant amount of time thinking about extraterrestrial life, his work revolutionized how the Universe operates and as a consequence, what environments are to be expected on other worlds.

Einstein

Life in the Universe and Where to find it

Otto Struve (1897 – 1963) was a prolific writer, tireless researcher and member of a prominent family of astronomers. He served as director of the Yerkes, McDonald, Leuschner and and National Radio Astronomy Observatories. He hired Novel prize winning researchers while at these facilities. His work focussed on binary and variable stars, stellar rotation and interstellar matter. He wrote over 900 journal articles and books on these subjects.

He was one of the few eminent astronomers to publicly admit that he believed in extraterrestrial intelligence, and that is was abundant in the Universe. As a result he was an early proponent of the search for extraterrestrial life. His beliefs on this subject were a result of his work on slow-rotating stars. He found that our sun as well as many other stars spin at a much lower rate than was predicted by early theories on stellar evolution. He thought that the reason for this was that they were surrounded by planetary systems and thus slowed by the loss of angular momentum. He estimated the number of planets around these types of stars at around 50 billion and reflecting on how many may harbor intelligent life he wrote:

"An intrinsically improbable event may become highly probable if the number of events is very great. ... it is probable that a good many of the billions of planets in the Milky Way support intelligent forms of life. To me this conclusion is of great philosophical interest. I believe that science has reached the point where it is necessary to take into account the action of intelligent beings, in addition to the classical laws of physics."

Struve

Enrico Fermi (1901 – 1954) was a famous Italian physicist that immigrated to the United States and did the first successful work on a nuclear reactor in Chicago. He also worked on quantum theory, nuclear and particle physics and statistical mechanics. In 1938 he was awarded the Nobel Prize in Physics for his work on induced radioactivity. Although he did not work at Los Alamos, he did visit there and was instrumental in the development of the atomic weapons programs.

To give you an idea of his intellectual prowess, when he was 17 he wrote an essay for an entrance examination for a particular collet entitled *Characteristics of Sound,* where he derived and solved Fourier equations based on partial differential equations for waves on a string. The examining professor interviewed Fermi and concluded that his work would have been commendable even for a doctoral degree candidate.

He also has become famous in cosmology circles for creating the "Fermi Paradox," which he stated "This contradiction or proposition is this: that with the billions and billions of star systems in the universe, one would think that intelligent life would have contacted our civilization by now." In other words:

"Where are they?"

This question was asked as an informal discussion on the topic where some of the first searches for extraterrestrial life had been performed. No indication of alien life had been (at the time of the discussion) or ever has been forthcoming. As a result this question is hard to answer.

Fermi

George Gamow (1904 – 1968) was a Russian born theoretical physicist and cosmologist. His areas of research included star formation, stellar nucleosynthesis, Big Bang nucleosynthesis, the cosmic microwave background, nucleocosmogenesis and genetics. He was particularly interested in the processes of stellar evolution and the early history of the solar system. In 1948 he developed equations for the mass and radius of a primordial galaxy which typically contains about one hundred billion stars, each with a mass comparable with that of the sun.

Gamow wrote an important paper on cosmogony named "The Origin of Chemical Elements" in 1948. This paper discussed the levels of Hydrogen and Helium in the universe as a consequence of the reactions that occurred during the Big Bang. Elements heavier than Helium were not addressed. Fred Hoyle later wrote on this issue. Gamow also estimated what the strength of the Cosmic Microwave Background (CMB) at 5 degrees above absolute zero (K). The true value turned out to be around 2.7 degrees K.

Gamow

Life in the Universe and Where to find it

John Kraus (1910 – 2004) was an American physicist and contributed significantly in the fields of antenna design and radio astronomy. His books on these subjects are still mainstays of University courses. He invented a multitude of antennas, namely the helical, corner reflector some of the largest and most successful radio telescopes every built. His surveys of the radio universe are known as the Ohio Sky Survey and found thousands of radio sources in a time when only a handful were known.

After he finished his doctorate, Kraus worked on nuclear physics helping to design a 100 ton cyclotron and during the second world war worked on degaussing large naval ships to minimize their susceptibility to proximity mines. He wrote not only textbooks on electro-magnetics, radio astronomy and antenna design, but wrote on his work designing and building very large radio telescopes.

At Ohio State where Kraus taught, the "Big Ear" was constructed, which is a combination of parabola on one side and a planar reflector on the other side both of which were 100s of feet wide. This is a transit instrument and has a feed point centered in a small structure in the center. When it was completed, they simply turned on the chart recorder and "started taking data". New radio sources were instantly discovered and one particular one called the "Wow" signal was an example of periodic large pules of radio radiation caught when an antenna is looking in the right direction at the right time. The advent of much larger telescopes lowers the probability of finding more of these signals as they necessarily have smaller fields of view.

In 1958, Kraus used the signals from WWV (the shortwave timing signals) to track the disintegration of the Russian satellite Sputnik. He explored the fact that the lower the satellite's orbit became the more it would ionize the atmosphere and thus reflect the WWV signal. As a result of this research, the idea of meteor scatter communications came into being. Today, this technique is used for commercial and military applications, transferring data at high rates of speed when meteorites (some 3 billion per day, mostly dust sized) strike the atmosphere.

Kraus thus contributed significantly to the development of radio telescopes and to the instruments of the eventual discovery of extra terrestrial life. His text book on radio astronomy contains details on high resolution telescopes as well as spectroscopic radio astronomy techniques.

Kraus

Life in the Universe and Where to find it

Fred Hoyle (1915 – 2001) wrote on the concept of nucleosynthesis in stars in 1946. These writings explained elements heavier than helium in the universe and showed that critical elements such as carbon could be generated in stars and then incorporated in other stars and planets when the first star came to its end of life. He also theorized that rarer heavy elements could be formed in supernovas. Today we recognize that most of these heavy elements like gold, platinum and silver did indeed come from supernovas and that all that we are made of is essentially star dust.

In later years he promoted the theory that life evolved in space and spread throughout the universe via panspermia and that the evolution on earth is driven by a steady influx of viruses arriving with comets.

This theory competes with the abiogenesis theory as the fundamental origin of life on Earth.

Hoyle

Life in the Universe and Where to find it

I. A. Shklovskii (1916 - 1985) was an astrophysicist and radio astronomer who proposed that cosmic rays from supernova explosions within 300 light years of the sun could have been responsible for some of the mass extinctions of life on earth. He made significant contributions to radio astronomy and wrote a book with Carl Sagan that argued for serious consideration of "paleocontact" with extraterrestrials in the early historical era, and for examination of myths and religious lore for evidence of such contact.

The book he co-wrote with Sagan is titled *Intelligence Life in the Universe.* Shklovskii is also the author of an important book in Radio Astronomy called *Radio Astronomy.* He also was the chief scientist for several of the most prominent radio telescope institutions in Russia.

Shklovskii

41

Life in the Universe and Where to find it

Robert Dicke (1916 – 1997) worked with Jim Peebles and other to re-predict the cosmic background radiation (CMB) left over from the Big Bang and designed the appropriate equipment to directly test the theory. Arno Penzias and Robert Wilson, working on a similar apparatus in 1965 used a receiver setup designed by Dicke to first make the confirmation of the CMB a few miles away from Dicke's laboratory at Princeton. This was an important example of how concepts in theoretical physics could be directly tested.

Dicke is famous for designing a special receiver that negates the changes in sensitive amplifier gain and noise levels. When measurements are made in radio astronomy, the levels being recorded are often very low. As a consequence the drifting of the low noise amplifier (LNA) which is directly attached to the antenna can be compared to a stable noise source and differentially corrected. This design was extremely important to radio astronomers and is used in its original form or a close variant in today's modern instruments.

Dicke

The Antenna that detected the Cosmic Background Radiation
at Holmdel, New Jersey

Freeman Dyson's (1923 -) initial career was working on the first atomic bomb at Los Alamos.. To get to his job, he needed to drive from Chicago to New Mexico. During this trip he was asked to take along another passenger, also headed for the laboratory. It turned out to be Richard

Feynman. As they both turned out to be some of the finest scientists in the World, one can only image the conversations that transpired during this drive. Dyson, along with a plethora of papers on the subject of astrophysics, also has speculated about extraterrestrial life. The Dyson sphere is named after him, as he speculated that an advanced civilization would build a sphere around a star to completely capture all of the energy radiating from it. The wasted heat from such a structure would radiate in the infrared portion of the electromagnetic spectrum. Therefore one method of searching for life in the Galaxy would be to pay particular attention to infrared astronomy.

Dyson suggested that very advanced civilizations may have such powerful technologies that they can engage in engineering efforts on a planetary or stellar scale. He visualized civilizations rebuilding their planetary systems to create additional habitable planets, planetoids, or huge orbiting space stations and thus provide more *lebensraum* (habitat or living space). Noting that this would increase the surface area at which radiation was occurring at temperatures on the order of 300 ° K, he suggests that such civilizations might be detectable as a result of the excess radiation in the 10 micron range, and concluded that a rational approach to the search for extraterrestrial intelligence is indistinguishable from an expanded program in infrared astronomy. [2]

Dyson

Carl Sagan (1934 – 1996) was an astronomer, cosmologist, author and science popularizer who published more than 600 scientific papers and authored more than 20 books.

He promoted the Search for Extra-Terrestrial Intelligence (SETI) and was a founder of the Planetary Society. Sagan devoted his life to research on cosmological issues and wrote many books on the subject, one of which turned into the movie "Contact" which popularized his notion of what extraterrestrial life would be like.

Sagan urged scientists to listen with radio telescopes for evidence of life in our galaxy and worked with many Nobel winning scientists to promote the idea. He also worked with Frank Drake, using the "Drake" equation to suggest the existence of a large number of extraterrestrial were possible but because of the Fermi paradox, these civilization had destroyed themselves quickly. The solution to such a problem, he thought, was to become a spacefaring species. He also, with help from Drake, created a message that was transmitted to the stars with the large 300 meter Arecibo radio telescope. This message was a binary encoded image that showed the existence of life on the third planet of our Solar System.

Sagan

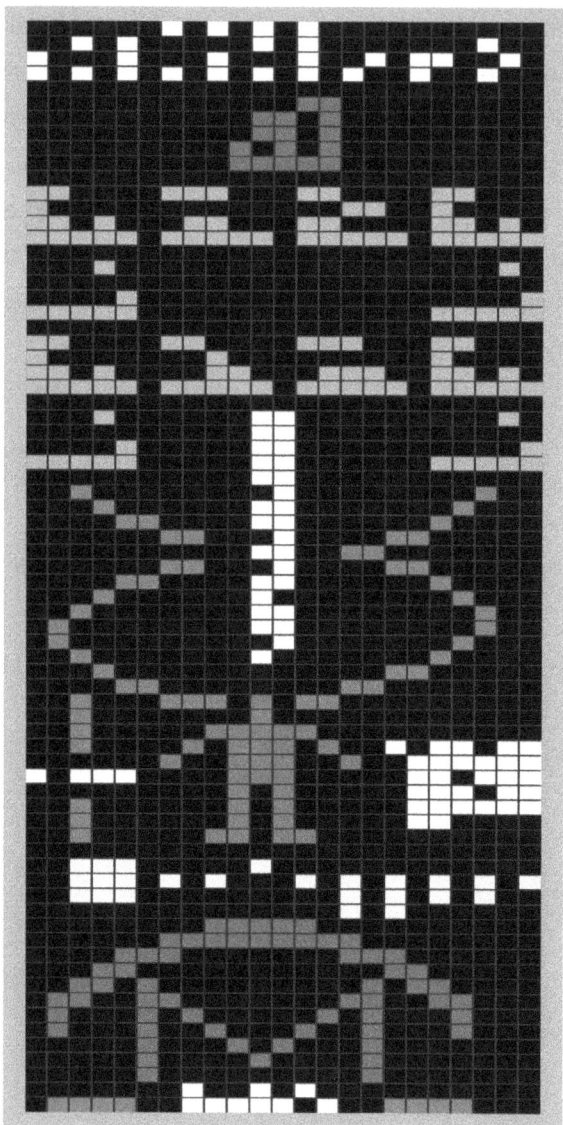

Binary Message sent to the stars designed by Sagan and Drake

Life in the Universe and Where to find it

Frank Drake (1930 -) considered the possibility of life in the Universe at a very early age. He attended Cornell University and began studying astronomy. Drake attended a lecture by Otto Struve at Cornell that motivated him to study the subject in greater detail. After he served briefly in the Navy, Drake went to graduate school at Harvard and studied radio astronomy there in earnest.

After he graduated from Harvard, Drake worked at the National Radio Astronomy Observatory in Green Bank, West Virginia. In addition he worked at the Jet Propulsion Laboratories in Pasadena, California. Initially, he studied the radio emissions from Jupiter and made important measurements which detected the presence of an Ionosphere and magnetosphere on the planet.

He was additionally interested in pulsars and did significant research in Arecibo, Puerto Rico at the 1000 foot radio telescope. This is the same telescope that he and Sagan eventually transmitted the binary coded message to the stars.

Later he again worked with Carl Sagan and designed the plaque that was affixed to the Pioneer 10 satellite that flew by Jupiter. Later he worked on the Voyager Golden record, which also flew to Jupiter.

Currently he is involved in the "The Carl Sagan Center for the Study of life in the Universe".

One of his most important contributions has been what is referred to as the "Drake Equation", which is a formula to calculate the number of civilizations in our galaxy which might be able to communicate with us.

The Drake equation states that:

$$N = R^* \cdot f_p \cdot n_e \cdot f_\ell \cdot f_i \cdot f_c \cdot L$$

where:

N = the number of civilizations in our galaxy with which communication might be possible;
and

R^* = the average rate of star formation per year in our galaxy
f_p = the fraction of those stars that have planets
n_e = the average number of planets that can potentially support life per star that has planets
f_ℓ = the fraction of the above that actually go on to develop life at some point
f_i = the fraction of the above that actually go on to develop intelligent life
f_c = the fraction of civilizations that develop a technology that releases detectable signs of their existence into space
L = the length of time for which such civilizations release detectable signals into space.

Alternative expression

The number of stars in the galaxy now, N^*, is related to the star formation rate R^* by

$$N^* = \int_0^{T_g} R^*(t)\,dt,$$

where *Tg* = the age of the galaxy. Assuming for simplicity that *R* is constant, then and the Drake equation can be rewritten into an alternate form phrased in terms of the more easily observable value, *N*

$$N = N^* \cdot f_p \cdot n_e \cdot f_\ell \cdot f_i \cdot f_c \cdot L/T_g$$

R factor

One can question why the number of civilizations should be proportional to the star formation rate, though this makes technical sense. (The product of all the terms except *L* tells how many new communicating civilizations are born each year. Then you multiply by the lifetime to get the expected number. For example, if an average of 0.01 new civilizations are born each year, and they each last 500 years on the average, then on the average 5 will exist at any time.) The original Drake Equation can be extended to a more realistic model, where the equation uses not the number of stars that are forming now, but those that were forming several billion years ago. The alternate formulation, in terms of the number of stars in the galaxy, is easier to explain and understand, but implicitly assumes the star formation rate is constant over the life of the galaxy.

Criticism of the Drake equation follows mostly from the observation that several terms in the equation are largely or entirely based on conjecture. Thus the equation cannot be used to draw firm conclusions of any kind. As T.J. Nelson states:

"The Drake equation consists of a large number of probabilities multiplied together. Since each factor is guaranteed to be somewhere between 0 and 1, the result is also guaranteed to be a reasonable-looking number between 0 and 1. Unfortunately, all the probabilities are completely unknown, making the result worse than useless."

Likewise, in a 2003 lecture at Caltech, Michael Crichton a science fiction author, stated:

"The problem, of course, is that none of the terms can be known, and most cannot even be estimated. The only way to work the equation is to fill in with guesses. [...] As a result, the Drake equation can have any value from "billions and billions" to zero. An expression that can mean anything means nothing. Speaking precisely, the Drake equation is literally meaningless..."

Another objection is that the very form of the Drake equation assumes that civilizations arise and then die out within their original solar systems. If interstellar colonization is possible, then this assumption is invalid, and the equations of population dynamics would apply instead.

One reply to such criticisms is that even though the Drake equation currently involves speculation about unmeasured parameters, it was not meant to be science, but intended as a way to stimulate dialogue on these topics. Then the focus becomes how to proceed experimentally. Indeed, Drake originally formulated the equation merely as an agenda for discussion at the Green Bank conference. [1]

Drake

Roger Penrose (1931 -) is a British mathematician and cosmologist. He has made several significant contributions to the field including the idea that singularities (like black holes) could be formed by the gravitational collapse of immense dying stars. He has written extensively on artificial intelligence as well as the fundamental laws of nature. His work has given insight into the details of black holes and the nature of the physical universe.

Penrose continues to teach at Cambridge and edits the *Journal of Cosmology*. He also edited the book *Consciousness in the Universe*.

Penrose

Stephen Hawking (1942 -) is a theoretical cosmologist from England and holder the the Isaac Newton Chair at Cambridge. In addition he is particularly interested in quantum gravity. He and his colleague, Roger Penrose used Einstein's theory of general relativity to predict the existence of gravitational singularities in space time, such as a black hole. He also was able to predict the condition just after the Big Bang where small black holes were created, emitting gamma rays and subatomic particles.

He has also worked on cosmic inflation, quantum cosmology, quantum entanglement, entropy, the arrow of time, string theory, space-time foam, supergravity, gravitational radiation and wormholes.

As far as alien life is concerned he says, "To my mathematical brain, the numbers alone make thinking about aliens perfectly rational. The real challenge is to work out what aliens might actually be like." In addition, he believes alien life not only certainly exists on planets but perhaps in other places, like within stars or even floating in outer space. He also warns of contacting these other worlds and reminds us, "If aliens visit us, the outcome would be much as when Columbus landed in America, which didn't turn out well for the Native Americans."

He advocates that we should avoid trying to contact other life forms. [1].

Hawking

Robert Zubrin (1952 -) is an American aerospace engineer and author, best known for his advocacy of the manned exploration of Mars. He was the driving force behind Mars Direct—a proposal intended to produce significant reductions in the cost and complexity of such a mission. The key idea was to use the Martian Atmosphere to produce oxygen, water, and rocket propulsion for the surface stay and return journey. A modified version of the plan was subsequently adopted by NASA as their "design reference mission".

55

First a spacecraft would land on Mars, create fuel while it waited for a second spacecraft to land with astronauts.

Zubrin holds a B.A. In Mathematics from the University of Rochester (1974), and a masters degree in Aeronautics and Astronautics, a masters degree in Nuclear Engineering and a Ph.D. in Nuclear Engineering — all from the University of Washington. He has developed a number of concepts for space propulsion and exploration, and is the author of over 200 technical and non-technical papers and seven books. He is also President of both the Mars Society and Pioneer Astronautics, a private company that does research and development on innovative aerospace technologies. Zubrin is the co-inventor on a U.S. design patent and a U.S. utility patent on a hybrid rocket/airplane, and on a U.S. utility patent on an oxygen supply system. He was awarded his first patent at age 20 in 1972 for Three Player Chess. His inventions also include the nuclear salt water rocket.

One of his books, Entering Space: Creating a Spacefaring Civilization (1999) discusses the advent of stellar exploration. This book discusses exploration and colonization of Mars, and moves to a more futuristic look at humanity's possible colonization of the solar system and the feasibility of interstellar travel with known physics. [1]

Fundamentally, Zubrin leaves us with the impression that we could achieve our goals of the exploration of Mars and the stars without reasonable risk. With current technology all that is required is a commitment. "We could go now."

Zubrin

Science Fiction Writers

Space is indifferent to what we do; it has
no feeling, no design, no interest in whether
or not we grapple with it. But we cannot
be indifferent to space, because the grand,
slow march of intelligence has brought us,
in our generation, to a point from which we
can explore and understand and utilize it.
To turn back now would be to deny our
history, our capabilities.

~ *James A. Michener*

Introduction

Science Fiction literature throughout the ages has opened the bounds of creativity and thus allowed for some very "far out" stories. In combination with reasonable accurate science, writers started to combine new ideas with what was currently understood about the Universe. The result was science fiction which captured and continues to capture the imaginations of millions of readers. There is something to be said about this combination as it relates to our curiosity and expectations. Humans tend to "lean" in one direction as they progress through time. In other words there seems to be an inherent need to explore, to look up at the stars and wonder what is out there and in the case of science fiction, take a guess as to what to expect. Many books have been written on this subject and several of those have been made into visual form by making movies. The overwhelming interest the public has in the movies and how these movies have effected us is remarkable. Take for instance, the TV episodes of *Star Trek* where many cultural and political ideas were played out in the future. When the producer Gene Roddenbury came up with the issue of multi-racial, multi-cultural crews, the viewing public had not experienced this before. He also produced shows based portraying omniscience, prejudice, sentient alternative life forms and galactic exploration. Movies have followed the TV series and other movies, notably *Star Wars* and *Contact* have entertained at least a billion people on this Earth. This is a consequence of the resonance between human beings and the ideas of the science fiction authors.

Although the last many pages have been focused on philosophers and scientists, it is important to include several of the most influential science fiction writers as they have fanned the flames of interest in so many of us.

Life in the Universe and Where to find it

Jules Verne (1828 – 1905) might be considered the first science fiction author as he wrote several amazing books which at the time were exploring completely new territory. He is best known for *Twenty Thousand Leagues Under the Sea* (1870) were he introduces Captain Nemo and his submarine Nautilus, exploring the depths of all of the oceans. The story starts out as a search for a giant sea monster which is attacking ships. This turns out to be the Nautilus, which is a large submarine running on electric power. This concept was of course way ahead of its time. Verne also introduces other interesting ideas in *A Journey to the Center of the Earth* (1864) and *Around the World in Eighty Days* (1873). In these books and others he explored air travel, space travel and travel through the Earth.

In his book *Around the Moon*, Verne discussed the possibility of life on the Moon, but concluded that it had none.

Interestingly, he also wrote the novel *Paris in the 20th Century* where he described a man who lives in a world of glass skyscrapers, high-speed trains, gas powered automobiles, calculators and a worldwide communications network. This person however cannot find happiness and comes to a tragic end. This book was written in 1863 but the publication was delayed due to its pessimistic outlook. Verne placed the manuscript in a safe where it remained until 1989, then published in 1994.

Jules Verne

H.G. Wells (1866 – 1946) was a very creative writer whose books became legends and in many cases, movies. *The Time Machine* portrays an inventor who discovers the means to project himself into the future and eventually back again. During his foray in the future he discovers that Earth has changed significantly with a society of beings living underground. They emerge periodically to harass or abduct the remnants of a human society living above ground. This book formulated the idea of time travel and looked into an apocalyptic future. His other books like *The Island of Doctor Moreau, The Invisible Man, The War of the Worlds, When the Sleeper Wakes and The First Men in the Moon* are equally inventive. The *War of the Worlds* discussed an interesting topic of what will happen to when life travels to other worlds and encounters disease and other biological organisms.

Wells believed in a more organized society which led him to write several Utopian novels.

As far as extraterrestrial life is concerned, in his book *The War of the Worlds* (1898), Wells depicted an invasion of aliens from Mars fleeing their world's destruction. He was influenced by the work of Lowell with regards to the possibility of life on Mars. This particular book was depicted in a radio show by Orson Wells in the late 1930s. The idea was believable enough to cause serious panic in the United States. Both Wellses became very well known after this event. Many of his books have been adapted to the movies.

The idea of time travel was revolutionary at the time of his book *The Time Machine*, since then many discussion have been had overt the subject. Many cosmologists have also tackled the subject and some have speculated that this once fictional topic might be feasible by using the effect of speed close to that of light and / or the effects of extreme gravity near black holes. This is one of many examples of ideas from fiction being taken seriously in the scientific community.

Life in the Universe and Where to find it

H. G. Wells

Ray Bradbury (1920 -) is an American science fiction author who has also written in the areas of fantasy, horror and mystery. He is known for his novels *Fahrenheit 451* (1953), *The Martian Chronicles* (1950) and *The Illustrated Man* (1951). Many of Bradbury's books have been made into movies.

The Martian Chronicles is a series of short stories that chronicle the colonization of Mars by humans escaping Earth and its atomic devastation from wars. Conflicts arise from the original Martians and the new colonists.

Life in the Universe and Where to find it

Fahrenheit 451, named after the temperature at which books catch fire is a dystopian novel about a firemen whose job it is to burn books. American society is portrayed as hedonistic where critical thought through reading is outlawed.

The Illustrated Man is a book of eighteen short stories which although not directly connected, have a recurring theme which is the conflict of the cold mechanics of technology and the psychology of people. These stories are tied together by the concept of a vagrant with a tattooed body allegedly created by a woman from the future, each tattoo tells a different story and a viewer can be lost by looking at them.

Bradbury

Life in the Universe and Where to find it

Isaac Asimov (1920 – 1992) was a professor of biochemistry and American author who rose to popularity after his *Foundation* series of books were published. Two other series of books, *Galactic Empire* and *Robot* created what he called a "future history." He also wrote many short stories including *Nightfall*.

The *Foundation* series of books revolves around the experiences of a psychologist who spends his life developing a branch of mathematics called "psychohistory." Using the laws of mass action it predicts the future on a large scale. The operating principle is that the behavior of a mass of people is predictable, especially if the mass is very large, equal to the population of the galaxy.

The *Galactic Empire* series of stories introduces many new concepts like hyperdrive, blaster pistols, neuronic ships and the visi-sonor. The whole series portrayed Asimov's idea of the future.

The *Robot* series is a set of short stories that include the concept of a positronic brain and the "Three Laws of Robots" which is a central theme in such books as *I, Robot* which became a movie in 2004. The *Robot* series is about space exploration, robot/human interactions and morality. *Bicentennial Man* was a story about a robot trying to become human where it discovers that the only way to do so is to become mortal. The robot is driven by the need to be human.

Life in the Universe and Where to find it

Asimov

Arthur C. Clarke (1917 – 2008) was a prolific British science fiction author who also discovered such things as the geostationary orbit (sometimes referred to as the "Clarke" orbit) used extensively today, this is where satellites are placed so they move around the planet at the same rate as the planet's rotation period, hence becoming "stationary" in the sky as viewed from the ground. He wrote several series of books including *2001 Space Odyssey* and the stories about *Rama*, a robotic ship that came to visit our solar system. He was knighted by Queen Elizabeth II in 1998 and was the chairman of the British Interplanetary Society.

He had the unique capacity to bring the reader in his world, which was made up of reasonable technology, nothing that would seem implausible, move us deeper into an unknown place as the reader followed along, and into a new thought provoking realm. Readers and viewers of his work were many times transformed into allowing their minds to speculate about the unique ideas Clarke introduced. These included a recurrent theme of meeting life that was vastly advanced compared to ours and how "magical" it could appear.

Many of his books turned into movies and ultimately showed us how different alien life will be.

Clarke

The Concept of UFOs

For several decades now there have been thousands of reports of Unidentified Flying Objects (UFOs). The most credible reports might be from pilots, who have witnessed unexplained phenomenon and described them. To date, no sighting has been confirmed as extraterrestrial. No physical evidence has been confirmed as extraterrestrial in origin. What seems to be interesting is that the sightings have occurred overwhelmingly in the United States.

There have also been numerous ruses depicted as reliable. Most of these have been exposed but one wonders how many others have not.

The classified testing of aerial vehicles has been reported as UFOs in many cases. One would not expect governments to discuss in detail their research and development projects. Also, the existence of high power laser weapons and high energy focused energy (not necessarily optical) beams will not be advertised.

Considering how long it would take to travel from even the closest star systems, one wonders why these objects are so small.

Is it possible that there is a psychological component with UFO sightings? Does the human race actually want to be visited and is looking for signs?

The First Sighting

On June 24, 1947, an amateur pilot named Kenneth Arnold was flying a small plane near Mount Rainier in Washington state when he saw something extraordinarily strange. Directly to his left, about 20 to 25 miles north of him and at the same altitude, a chain of nine objects shot across the sky, glinting in the sun as they traveled. By comparing their size to that of a distant airplane, Arnold gauged the objects to be about 45 to 50 feet wide. They flew between two mountains spaced 50 miles apart in just 1 minute, 42 seconds, he observed, implying an astonishing speed of 1,700 miles per hour, or three times faster than any manned aircraft of the era. However, as if controlled, the flying objects seemed to dip and swerve around obstacles in the terrain. When the objects faded into the distance, Arnold flew to Yakima, Wash., landed and immediately told the airport staff of the UFO he had spotted. The next day, he was interviewed by reporters from a local newspaper and the story spread like wildfire across the nation. [1]

From this point on, thousands of sightings have been reported. The US Air Force got involved at one point writing a report called *Project Blue Book*. There are still reports coming in with varying amounts of evidence. Many of these reports are in proximity to Area 51 in Nevada, a proving ground for new aircraft.

Most of the sightings have been attributed to meteorological phenomena, new aircraft types and hoaxes. For the most part there is a lack of credibility in the reports and new ones are typically viewed with skepticism. There are however a small amount of sightings that are documented by very credible people. One needs to ask the question: "Why don't they stop?"

Conclusion

Considering life in outer space has been on the human mind for

Life in the Universe and Where to find it

thousands of years.

Philosophers, Scientists and Science Fiction authors have made significant efforts in understanding and predicting the inevitable discovery of life in the Universe.

Other, non-professional activities reflect the human interest in finding other life in the Cosmos.

Chapter 3 - Radio Astronomy

The 100 meter Green Bank Telescope (GBT)

'Thus the explorations of space end on a note of uncertainty. And necessarily so. We are, by definition, in the very center of the observable region. We know our immediate neighborhood rather intimately. With increasing distance, our knowledge fades, and fades rapidly. Eventually, we reach the dim boundary – the utmost limits of our telescopes. There, we measure shadows, and we search among ghostly errors of measurement for landmarks that are scarcely more substantial. The search will continue. Not

Life in the Universe and Where to find it

until the empirical resources are exhausted need we pass on to the dreamy realms of speculation.'

-Edwin Hubble

Fundamentals

Radio Telescopes are essentially very large antennas connected to very sensitive receivers and then to sophisticated data analysis software These instruments have the ability to look deep into the Universe in great detail to measure what is in the lower portion of the electromagnetic spectrum, from below the AM band to just below the Infrared band.

Some of the most interesting and exciting astrophysical phenomenon have been accomplished by radio astronomers using these tools. Simply put, radio telescopes are designed for the highest gain, highest resolution and highest sensitivity possible. These qualities are necessary for the further discovery of interesting planetary, galactic and extra-galactic details. The very origin of the Universe with the 2.7 Degrees K afterglow from the Big Bang was discovered by radio astronomers.

To date the discovery of noise storms on Jupiter, thermal radiation from the Milky Way, Pulsars, Quasars, emissions from ionized gas, masers, stellar jets and a host of other physical phenomena has been accomplished by these special telescopes. In addition thermal radiation, synchrotron, spectral lines, absorption lines, Zeeman splitting, Doppler shifting are but some of the physical characteristics viewed by these instruments. The determination of the

72

shape of the Milky Way, its core and the motion of the arms were discovered by these telescopes as well.

Sensitivities of better than 10^{-26} watts per square meter is typical of these telescopes, with antenna gains beyond 100 dBi and beam widths (using interferometric techniques) in the micro arc-second range. Antenna arrays spanning continents coupled with antennas in space have increased the resolution of these telescopes to many times better than the best optical instrument (including the Hubble Space Telescope). These instruments are rarely encumbered by the atmosphere and can be used day or night.

A radio telescope in Holmdel, New Jersey verified the existence of the 2.7 degree Kelvin afterglow of the creation of the Universe. One in Puerto Rico is 1000' feet in diameter and has examined minute details of Pulsars, sent messages to our interstellar neighbors and starred in several movies.

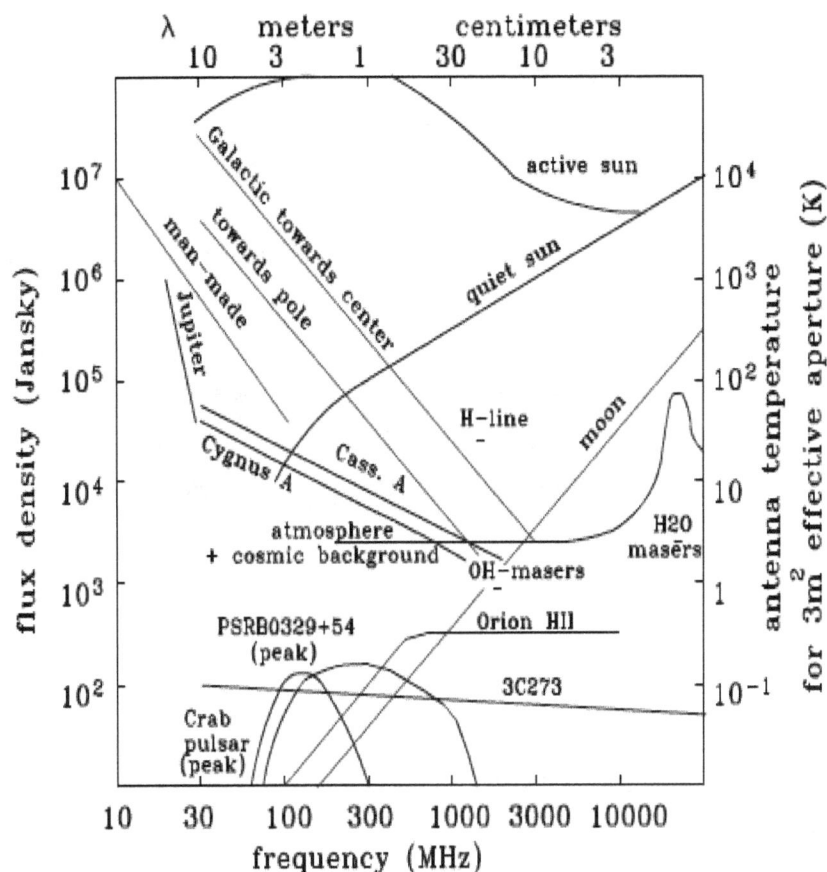

The Radio Astronomy Spectrum

History

Between 1861 and 1862, James Clerk Maxwell formulated a series of equations that described the workings of charges, magnetism and currents in a way that predicted the ability to create radio waves. Reformulated to a more useable form by Heaviside, these equations were used by Heinrich Hertz to create the first antennas and then the first communications from a distance, without wires.

Tesla, Marconi and Armstrong followed to create the first long distance short wave communications equipment. This equipment was used to send Morse code messages across oceans and from ships to shore.

Karl Jansky (1905 - 1950)

As a result of Tesla's, Marconi's and Armstrong's work, it soon became obvious that long distance communications at short and long wavelengths were subject to changes in the atmosphere and in particular, changes in the ionosphere caused major changes in the ability of communicate occur at certain wavelengths. In general, it was found that nighttime communications were better than daytime. Depending on the time of day or night, there was a maximum usable frequency that if exceeded, caused blackouts. Much research was being performed in this area in the laboratories to try to understand this phenomena. Karl Guthe Jansky was working at Bell Telephone laboratories during this period and was assigned the task of analyzing the variability of long distance communications. To do this he designed and built a directional antenna array known as a Bruce curtain. This consisted of a serpentine driven element in front of a reflector of similar dimensions. At 14 meters frequency, (approximately 21.4 MHz) this antenna was quite large and was supported by automobile tires from a Ford Model T on a concrete circular track. This antenna was movable in azimuth so that Jansky could analyze the positions and intensities of atmospheric noise over the entire sky. After several months of work, Jansky noticed that a faint hiss could consistently be heard in the direction of the Milky Way.

With these words, the field of radio astronomy was born:

"The data give for the coordinates of the region from which the disturbance comes, a right ascension of 18 hours and declination of -10 degrees."

This position corresponds to the center of the Milky Way. The radio receiver was tuned to 14 meters, in the middle of today's shortwave band. Jansky noticed that the radio source followed a sidereal day as apposed to an "Earth" day and so was able to place the source in the celestial sphere. The receiver recorded a high noise level, and as the earth turned the peak of the galactic noise was significant and moving at a sidereal rate.

As a consequence of his discovery, Jansky founded the field of Radio Astronomy and now has the honor of having his name used as the primary unit of measurement (flux density).

Jansky

Early Radio Astronomy Antennas

As mentioned above, Carl Jansky built the first antenna for use in characterizing shortwave propagation. In essence this antenna acted like a phased array with directional qualities. As a consequently, Jansky was able to point the antenna to large "spots" the sky and make measurements. Using rudimentary receiving equipment, he was able record atmospheric and exo-atmospheric radiation. This antenna was built in 1927 and is still intact and viewable at the National Astronomy Observatory in Greenbank, West Virginia. After the work of Jansky and the amateur radio pioneers came WWII where most radio amateurs were called into duty. During the war, the government strongly considered getting rid of the amateurs in favor of a radio operators solely for military use. This mandate was reversed based mainly on the effort of the first president of the ARRL (Amateur Radio Relay League). As a result, many new innovations from the amateur community including antenna designs were put into use.

Karl Jansky and the first Radio Telescope, a Bruce Aerial

In 1937, an engineer in Illinois named Grote Reber (also a Radio Amateur) had read the results of Jansky's research and wanted to analyze the galactic noise further. He constructed a 30 foot prime focus parabolic dish in his back yard and placed within the focal point. He also designed several receivers, one at 160 MHz and another at 3 GHz. The antenna's mount was an altitude over azimuth configuration and considering it has a circular beam (as apposed to a fan beam like Jansky's apparatus), Reber was able to map the "radio" universe in reasonable detail for the first time. Reber published several papers on his results in the *Astrophysical Journal* (by the way, the editor at that time, Otto Struve, concerned about the non standard language of Reber, took responsibility for the publications) and what resulted from Reber's efforts was the discovery of the remnants of a supernova remnant (Cassiopeia A), an extragalactic radio source (Cygnus A) as well as more definition of the radio emission from the galactic plane. Later, Reber was the first to publish a paper on radio emissions from our sun.

During WWII, radar operators in England, working at about 178 MHz noticed interference while scanning the skies at dawn and dusk. The emissions coincided with the position of the sun, just as Reber had discovered years before and as it rose during the day, and when out of the antenna beams, the   interference subsided. Scientists at the Cavendish radio astronomy group at Cambridge University were tasked with analyzing and hopefully nullifying the solar interference. They used search light reflectors initially and soon were able to pinpoint and track the solar radiation. It became apparent that the solar radio outbursts were not steady, solar "storms" were discovered and eventually the rise and fall of the number of storms over an 11 year period was verified to be consistent with the rise and fall of sun spots over the same period.

After the war, radar receivers and antennas (which were the recipients of significant improvements) were modified and used to make more accurate maps of the radio universe. In Australia, a group of researchers in Sydney also started work on several telescopes, including a "Mill's Cross" named after its inventor Bernard Mills, which discovered radio emissions from Jupiter. This antenna was constructed of a series of smaller antennas in a cross configuration to give the high resolution of a larger telescope at a significantly lower cost.

Back in England, the discovery of the all important 21cm line, inaugurated spectral radio astronomy. With this discovery came the mapping of the motions of the arms of the Milky Way (this due to doppler shift), as well as the existence of emission from our galactic neighbor, the Andromeda Galaxy.

In the 1950s and 1960s, several large telescopes were constructed including catenary types. Blackett, Lovell and others constructed a 218 foot diameter reflector, suspending wires on a ring of scaffolding poles and filling it in with 16 miles of wire. In addition, there was a 126 foot steel mast in the center holding the focal point antenna, this mast could be moved up to 15 degrees from the vertical to move the beam due to reflection, +/- 30 degrees. A significant number of discoveries and papers were obtained from this antenna including the 21 cm Hydrogen line emissions from the Andromeda Galaxy. Later even more sophisticated antennas were constructed, like the one below:

The Jodrell Bank 250 foot Radio Telescope

Located in Jodrell Bank, England, this 250' radio telescope was built from battleship components in 1957. The elevation pinions are actually gun turret components salvaged from a British WWII battleship.

In 1967, Jocelyn Bell of Cambridge, after analyzing 100s of feet of chart paper, noticed some curious anomalies in the records. The records were made from a new 81.5 MHz Mill's Cross telescope in England designed to examine quasars (or quasi stellar objects). Bell noticed that the recording rate of the chart paper was not fast enough to resolve all of the attributes of several of the sources recorded. Using oscilloscopes, faster chart speeds and audio techniques, it was discovered that these unique radio sources, now known as pulsars were in fact neutron stars rotating at very fast rates, some faster than 30 milliseconds per revolution. Confirmation was achieved through the gating of optical images at several optical observatories, including one at Flagstaff, Arizona. The gating was achieved by spinning a disk with a hole in it a rates slightly slower that the rate determined by Bell, the resultant image would slowly pulsate in intensity over a slower period.

In the U.S. The National Radio Astronomy Observatory was established and built several very large telescopes, including a 300 foot transit antenna, 140 foot high frequency antenna, an 85 foot telescope (used for one of the first extraterrestrial intelligence searches) and a large interferometer (or set of antennas) used eventually to design the Very Large Array in Socorro, New Mexico.

Using interferometric techniques derived from English and American efforts, The Very Long Baseline interferometer (VLBI) system was then developed. As a telescope gets larger, the beam width gets smaller and finer details of astrophysical phenomena can be observed. From a practical point of view, a single telescope become prohibitively expensive over very large diameters. The solution is to separate one or more individual telescopes to improve the resolving power. These are known as interferometers and are usually just large enough to couple the outputs of the separate receivers together via long cables or microwave links. In the 1970s, the development of very stable atomic oscillators and video tape recorders allowed the separation of telescopes to be expended to thousands of miles. Today there are orbiting elements of the VLBI network, allowing for resolutions in the micro arc-second range, vastly exceeding the resolution of optical telescopes.

Today, interferometers using millimeter and sub-millimeter wavelengths have been built in several countries. The advantage of these significantly higher frequency telescopes is the high resolution spectral imaging of interstellar molecules, gas clouds, and other high energy phenomena.

The Atacama millimeter and sub-millimeter radio telescope at 16,000 feet elevation, Northern Chile. Note curved struts on antenna to the left, this is done to minimized scattering off of feed positioning structures. This telescope is capable of mapping radio sources to 720 GHz.

The Radio Astronomy Equation

Generally speaking, the minimum flux sensitivity of a radio telescope can be described in the following equation:

$$\Delta S = \frac{(2 * k * Tsys)}{(Ae * \sqrt{(2 * \Delta T * \Delta V)})}$$

where:

S = Minimum Detectable Flux

k = Boltzmann's constant ($1.38 * 10^{-23}$)

Tsys = the system noise temperature (Antenna + Receiver), degrees K

Ae = the effective area of the telescopes in square meters

T = the observing time in seconds

V = the bandwidth of the received signal in Hz

Based on this formula, it should be obvious that the lower the system temperature, (created predominately by the first low noise amplifier after the feed antenna, added to a combination of the cable losses and "back end" losses of the receiver), the better. Also the larger the aperture, the longer the integration time, and the wider the bandwidth, sensitivity will increase.

A consequence of a large aperture however is lower beam width, or the ability to discern finer details of astrophysical phenomena.

System Noise Temperature

The system noise temperature heavily influences the ability to detect faint radio astronomical objects. It can be calculated by understanding the characteristics of the antennas, low noise amplifiers and the local environment. The following formula should be evaluated and entered into the radio astronomy equation to fully understand the detectability of radio sources:

$$Tsys = \frac{Ta}{Lf} + \left(1 - \frac{1}{Lf}\right) * Tout + Tr$$

where:

Tsys = System Noise Temperature

Ta = Antenna Temperature at the "antenna port"

Lf = Front end loss (from antenna port to preamp input)

Tout = reference temperature (290 degrees K for

ambient)

Tr = Receiver noise temperature

Antenna Beam Movement

Mechanical

The most obvious way of moving a radio telescope beam around is with a moving structure, able to change elevation and azimuth as needed to measure source "temperature" or flux. This kind of structure enables tracking of a particular piece of sky to allow long term integration and thus improve sensitivity. Most modern radio telescopes employ this method, a few have equatorial mounts, whereby one axis points to the North pole (or South pole if below the Equator). The other axis or declination, is set once and does not need to be moved further, enabling the equatorial axis to move with the rotation of the earth and thus track a source. This style is rare for radio telescopes but very popular with optical types. The largest example of an equatorial mounted radio telescope is the the 140 foot antenna at the National Radio Astronomy Observatory (NRAO) in Green Bank, West Virginia. See the following picture of this famous telescope:

The NRAO 140 foot 20 GHz Equatorial Mounted Radio Telescope

Electrical

Using a multitude of antennas separated by 0.5 to 0.75 lambda (for the simplest case) in a particular pattern defines a phased array. In either one or two dimensional layout, the antenna will receive a signal efficiently orthogonal

87

from the plane of the aperture if all of the antenna elements are phased the same relative to a summing feed point.

If, however, these elements are phased sequentially, for instance by ten degree increments, the antenna beam will move off the orthogonal orientation and tilt towards the short end of the phase distribution. As a consequence, many beam positions can be created allowing a wide angle of view with a small beam size. This is the basic principal of many military radar systems and in radio astronomy has been used to great effectiveness.

These phase offsets were initially created by different lengths of coaxial lines and eventually, with electrically switched delay lines using relays, PIN diodes or other types of electronic switches.

In another manifestation, John Kraus discovered that if you place three helical antennas in a line, and add (in phase) their outputs, then physically spin the outer helixes, the resultant beam moves along the axis of the line. By adjusting the azimuth position of the outer elements to a particular angle, a stationary offset beam can be created. This is in essence changing the phase between antennas

Earth Rotation

Transit

If a radio telescope is set to a particular elevation and azimuth and set in place, the earth will of course continue moving and automatically scan the antenna beam across an arc coincident with a line of latitude. This

88

technique has been used often to allow a map of the radio universe to be built up by simply setting the azimuth to North or South, then adjusting the elevation in single beam widths every sidereal "day." Over a period of many days, a celestial radio map is thereby constructed.

Synthesis

By the 1960s, radio astronomers realized that if they placed several antennas in a two dimensional interferometric array (single antennas separated by many wavelengths) and tracked the astronomical source of interest, the movement of the earth would create a multitude of projected "baselines" (or lines between each element) and allow two dimensional imaging with great precision. These baselines change in length as the earth moves due to the fact that the relative orientation of the antenna array changes, looking from the radio source towards the array. This is the basic concept behind the VLA (Very Large Array, Socorro New Mexico) and the VLBA (Very Long Baseline Array), across the U.S. with elements in Europe and in space.

Elements

Radio Telescope elements are many time made up of many smaller antennas from microstrip patches to fully steerable parabolic dishes. The following examples show some of the more common types:

Example of an Interferometer Element in Australia

Yagis

These antennas have been used to form phased arrays, interferometers and other components of radio telescopes. Simple in construction and easy to reproduce, they are wideband and have reasonable gain.

Horns

Horn antenna have been instrumental in discovering major phenomena in our universe. The following horn discovered the Hydrogen emissions in our galaxy, in addition, this antenna found the doppler shift associated with moving clouds of Hydrogen gas, allowing astronomers to understand the movement of the Milky Way arms and associated components.

21 cm (1420 MHz) Horn that discovered the Hydrogen Line

Helixes

Helix type antennas have been an important part of radio astronomy history. An inherent feature is that they are circularly polarized and can receive signals unaffected by the earth's ionosphere, at frequencies below 1 GHz. Linear type antennas are effected due to the Faraday rotation effect. Banks of helixes were designed and build by John Kraus and made some of the earliest detailed maps of the radio sky.

The 96 element helix array, designed by John Kraus and responsible for the early detailed mapping of the radio universe.

Parabolics

The majority of radio telescopes are of the parabolic kind, including some interesting modifications of the generalized shape. Having shown several examples of large standard parabolas, in this section we shall show some of the variations on the theme that have contributed significantly to the study of the radio universe.

At Ohio State, the "Big Ear" was constructed, which is a combination of parabola (360' x 70') as seen on the left side and a planar reflector (340' x 100') as seen on the right side of the following photograph. This is a transit instrument and has a feed point near the right side centered in the little structure. When it was completed, they simply turned on the chart recorder and "started taking data". The amount of new radio sources recorded due the high sensitivity of this antenna allowed Ohio State to create a list of sources with several hundred entries. A similar design on a larger scale was constructed in Marseilles, France.

The Ohio State "Big Ear", designed by John Kraus and source of the Ohio State Radio Astronomy Survey.

Russian Ratan

This design is a parabolic section, much like the rim of a standard parabola, which focuses the radio waves to a central point for reception. The advantage is significant resolution but with the inability to track sources and steer over large angles.

The telescope in the photograph is in Russia which has very wide band capabilities and is being used for quasar and pulsar research.

The 600 meter diameter Ratan-600, outside of Moscow

Arrays

Mill's Cross

An un-filled aperture, made up of two lines of antenna elements, or Mill's cross has the resolution of a filled aperture of similar dimensions. Elements used in the construction of such arrays have been dipoles, smaller dishes, horns etc. with the predominant orientation of N-S for one array and E-W for the other array. Each array produces a fan beam which overlap orthogonally in the sky at one particular spot. These arrays can be phased to cover a significant amount of sky with this common crossover point. By adding

½ wavelength to the summed arrays in a periodic fashion. This cross over point will change in and out of phase and thus modulate the summed signal from both arrays (or blink). Making a differential measurement of this modulation allows the observer to measure the radio intensity of that small spot on the sky.

Pulsars were discovered with this technique in 1967 and many papers have been written from observations with this style of telescope.

Obviously building a cross was significantly cheaper than a full aperture dish, so they became very popular in the scientific community. Other forms of "aperture synthesis" became prevalent, notably American, Russian, Australian, English, and French astronomers who developed a vast assortment of similar antenna designs that researched such phenomena as quasars, pulsars, supernova remnants and Jupiter emissions.

Bernard Mill's Cross

Two element interferometer

A simple interferometer can be made up of two antennas, which can be single elements or arrays. The distance between the elements dictates the resolution of the array, the field of view of the interferometer is dictated by the elemental pattern. A series of fringes (where the radio sources become in and out of phase) is observed with the summed outputs of both elements. Examining the fringe details reveals the size of the radio source. As the distance between elements increases (as in a synthesis array), the fringes become smaller. This attribute continues until the radio source is resolved, it is at this point that the observer can determine the extent of the source

97

Multi element interferometers

Extending this concept to more than two elements allows for the creation of multiple baselines. A two element interferometer has one baseline, a three element has three baselines, a four element has six, etc. The baselines are between any two of the elements to create a "line" of measurement projected onto the sky. With multiple lines, a two dimensional "picture" of the radio source can be derived. The more elements, the better the clarity of the "picture".

The Very Large Array (or VLA) in Socorro, New Mexico is made up of 27 moveable 90 foot dishes in a Y configuration. These dishes are typically placed in one of three configurations, giving low, medium and high resolution depending on the requirements of the science being conducted. Each dish has several low noise receivers set to about 330,660,1420 and 1666 MHz. Also, there are S band, C band and X band frequency capabilities for use with the NASA Deep Space Network and the Arecibo 1,000 foot dish. Each baseline between each possible pair of dishes is sampled, correlated, and run through a plotting algorithm to produce very detailed maps of radio sources. There are 351 total baselines created by the 27 antenna array.

The Very Large Array in Socorro, New Mexico

The Very Long Baseline Array is made up of several 90 foot dishes, each with multiple feeds (10 bands) whose elements are placed from the Virgin Islands to California. Each element receives a signal, records it on wideband video tape, along with a reference signal derived from a very stable oscillator. These oscillators are typically Hydrogen Maser type or Cesium Beam with long term accuracies of one part in 10^{-15} seconds. As such all the elements although not connected by cable or microwave link, are synchronized very accurately. All of the combined baselines are sampled as with the VLA, and maps are produced with resolutions in the micro-arcsecond range, significantly better than any optical instrument. These arrays have discovered such phenomena as stellar jets (which initially appeared to be moving several times faster than light) and amazing details of the center of the Milky Way and the centers of other galaxies. Drs. Cohen, Shapiro, Kellerman, and Moran were all developers of this amazing instrument. A Japanese satellite with a large antenna, similar receivers and oscillators was recently flown to extend the baselines from several thousand miles to several tens of thousands of miles.

Coelestat

Invented by Russian radio astronomers in the 1950s, this antenna is made up of at least two elements back to back (for instance two dishes) aimed at two flat reflectors, the resolution like the previous interferometer, is dictated by the distance between the reflectors but the advantage is that a single receiver can be used without long cables and extra amplifiers.

Ring Array CSIRO Solar Heliostat

Located outside of Parkes, Australia, a ring of 15' diameter dishes has been constructed in a circle of approximately 500' in diameter. This ring of antennas was fed in phase to a central laboratory where there is an imaging apparatus and a series of receivers set to VHF and UHF frequencies. The small dishes tracked the sun and the outputs where (in one manifestation) connected via receivers to transducers on one end of a barrel of water. These transducers were arranged in the same relative position and order as the outside array. Another set of transducers were arranged in a X-Y arrangement

100

on the other end of the barrel and using the focusing properties of the water, were able to produce an image of the sun at several wavelengths. Using this apparatus, astronomers were able to watch ionized gases and highly magnetized loops move about the solar surface and above it. Eventually the water barrel was replaced by digital electronics capable of providing even higher resolution images. Several manifestations of this design have been built around the world specializing mainly in solar research.

Solar Radioheliograph in Australia

101

Life in the Universe and Where to find it

Lloyd's Mirror

This type of radio telescope is always found near a large body of water, typically the ocean. A reasonably large antenna is pointed to the East (normally) and the elevation is set to the horizon. The combination of the direct rays and the reflected rays from the water's surface of the radio source create an inexpensive interferometer. The field of view is somewhat restricted with this approach, however many high resolution maps have been made of the radio universe using this technique.

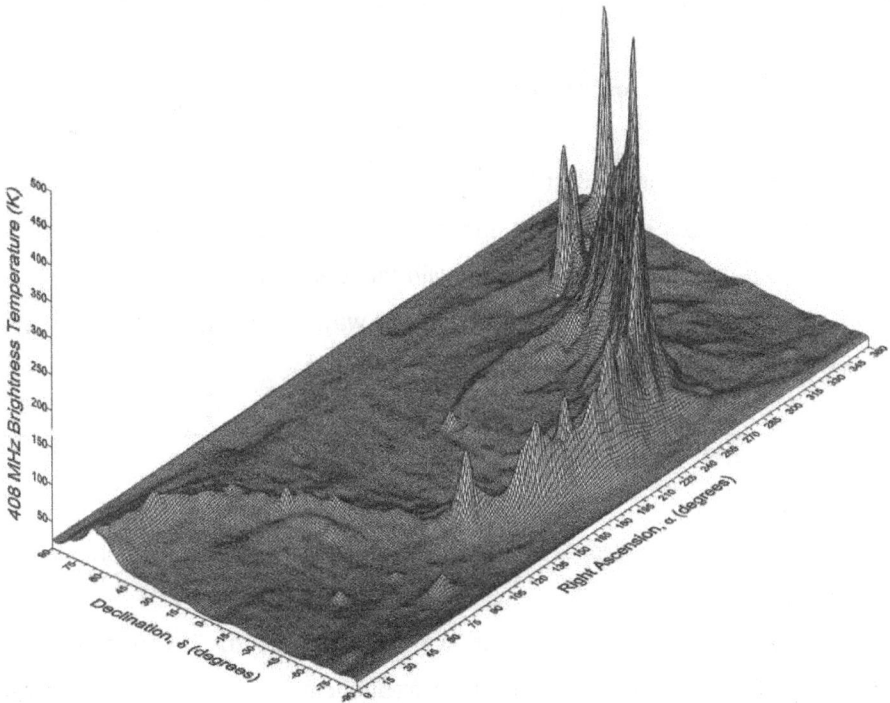

The Radio Sky at 408 MHz

102

Other Types

These antennas have been used to a lesser degree for radio astronomy purposes.

Conical

A large array of conical spirals was assembled over several acre at China Lake in the 60's for the purpose of bouncing signals off of the Sun's corona. The conical spirals allowed for very broad bandwidths.

Horn calibrators NRAO – Cass A

At Green Bank, West Virginia a horn antenna approximately 20 feet long is permanently placed at the same declination as the Cassiopeia supernova remnant. Data was taken with this transit instrument of the signal level on a daily basis for many years. In this way, the decay of power for this stellar remnant was observed.

VLF (Reber)

Grote Reber, a pioneering radio astronomer, moved to Tasmania in the 1970's to construct a very large series of arrays to explore the frequencies at and below 1 MHz. The research was difficult due to the fact that the ionosphere is typically opaque at these frequencies. On rare occasions, an "ionospheric lens" moves overhead whereby a significant amount of radio radiation can propagate through the atmosphere and at times focus and amplify the flux for easy detection on the ground.

The lower the frequency of observation, the more predominant the effects of Earth's ionosphere as mentioned above. The very lowest frequencies, from approximately 30 to 100 KHz, the effects of the magnetosphere can be observed. In fact there has been some research done trying to correlate the occurrence of a Gamma Ray Burster (GRB) to radio emissions at these very low frequencies.

Decametric (Radio Emission from Jupiter)

Starting in the 1950's, signals from Jupiter were observed using a Mill's Cross. Later arrays of wide band antennas, like log periodics, corner arrays with discone feeds and other wideband approaches were designed and built for this purpose. The radio bursts were categorized into at least five types with the predominant emissions coming from three distinct Jovian latitudes. The closest moon, Io also has an impact on the emissions as it goes through a plasma torus surrounding the great planet. The wide band emissions have unique characteristics that in some way emulate similar outbursts from the sun. This led to the theory that Jupiter was actually a proto-star and in fact produces more energy than is reflected from the sun. If the sun where to cease producing light for a moment, we would still be able to view Jupiter, as it glows due to its own energy output. The emissions can also be likened to a slowly rotating pulsar.

MRAS

The Mobile Radio Astronomy System (MRAS) was designed in the 1980s to address the need for Very Long Baseline Interferometry (VLBI) where a mobile radio telescope could enhance the quality of radio source mapping by moving a array telescope to optimize the UV plane (or sky map) coverage. One of the features of this telescope had the ability to integrate signals over the beamwidth (or field of view of all of the individual elements) or combine the elements as an interferometer to maximize resolution.   This feature is desirable when looking for chemical signatures in extra-solar planetary systems.

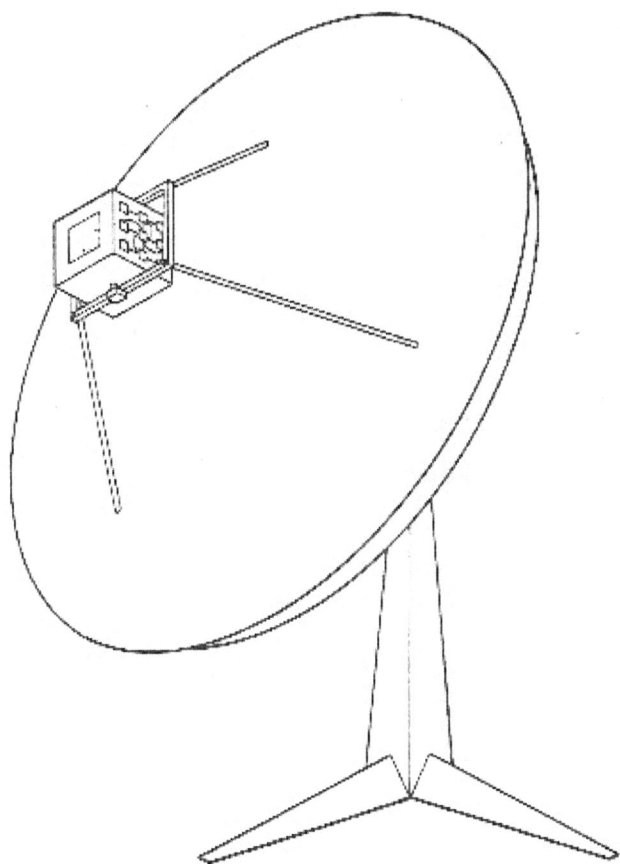

MRAS element

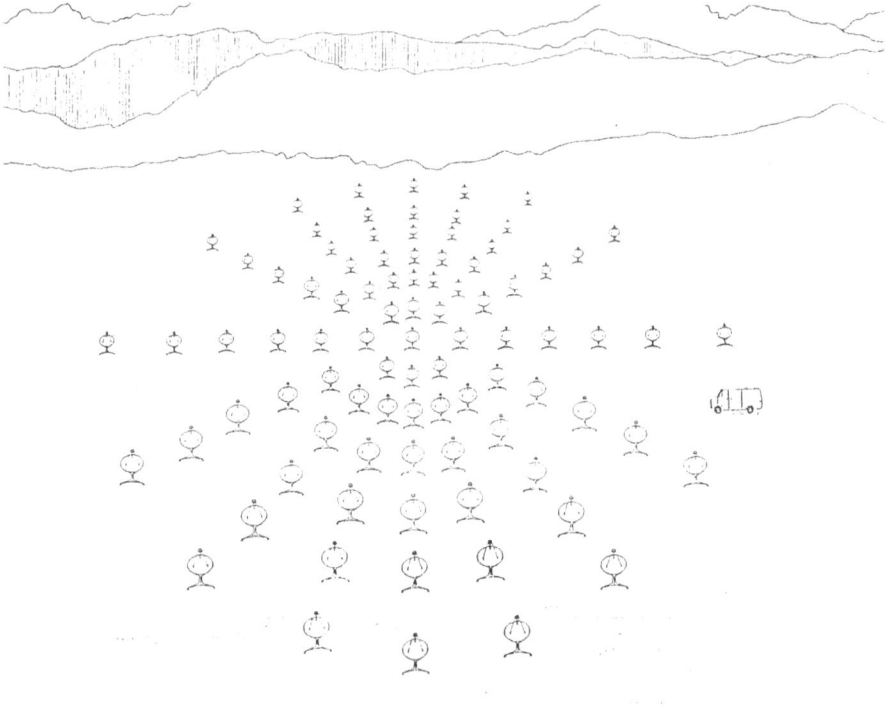

MRAS Array

LOFAR

The LOw Frequency ARray or LOFAR is a project conducted by the Westerbork Synthesis Array Observatory in the Netherlands. The array consists of a multitude (36 so far) of antenna sights place from 100 meters apart to 1500 kilometers apart in a roughly sparsed configuration. This array uses GPS and Rubidium Vapor clocks at each station to phase lock the received signals, perform first order data reduction and send the information via the Internet to a central processing facility. This facility has

107

supercomputers to perform correlation of the received signals from each station to all of the others. This is modeled after the VLBI (Very Long Baseline Array) currently in use in the United States at the National Radio Astronomy Observatory.

Each station has two sets of antennas, a low frequency portion that works from 10 to 90 MHz and a higher frequency portion that operates from 110 to 250 MHz. Two orthogonal polarizations are measured simultaneously.

A small portion to a full sky image can be obtained from this array, allowing for the ability to perform several useful scientific measurements. [1]

From the LOFAR website:

LOFAR will focus on six areas of research:

1. The Epoch of Reionisation - how did the first stars and black holes make the universe hot?
2. Extragalactic surveys - what is the history of star formation and black hole growth over cosmological time?
3. Transients and Pulsars - probing the extreme astrophysical environments that lead to transient bright bursts in the radio sky (see Anatomy of a Received Signal).
4. Cosmic rays - what is the origin of the most energetic particles in the universe?
5. Solar and space environment - mapping the structure of the solar wind, how it relates to solar bursts, and how it interacts with the Earth?
6. Cosmic Magnetism - what is the origin of the large-scale magnetic fields that pervade the universe?

The LOFAR low frequency antenna element

Modes of Operation:

Interferometric :
Single Source Imaging and Surveys

The central processor cross-correlates and averages the data streams from pairs of stations in order to compute interferometric visibilities. This is the basis of standard imaging and/or uv-plane (or sky map) analysis. The maximum time/frequency resolution is about 0.5 s / 0.7 KHz.

Tied Array:

Pulsars, Total Intensity time-series, Timing mode, Solar Monitoring Campaigns

The data streams of one or more stations are added without averaging on the sub-band level. This is the basis of the pulsar observing modes. The minimum integration time is 5 μs. The maximum spectral channel with is approximately 196 KHz.

Transient Buffer Board Use:

Ultra-High Energy Cosmic Rays, lightning detection.

The signal is monitored at station level by a triggering algorithm on the Transient Buffer Boards (TBBs). If triggered, the contents of the boards are frozen and sent out for further processing. This is the primary mode for cosmic ray observations, or lightning detection. The minimum integration time is 5 ns, and the maximum bandwidth 100 MHz. The transient buffer board can dump at most one second of data per event.

Direct Data Storage:

All sky imaging, spectrum monitoring, intra-station baselines, local transient buffer board experiments.

In this mode, the station will *not* send its data back to central processor for correlation. Correlation (if needed) will be done with the station correlator. The data will be directly stored and can be analyzed at the archive cluster, or at some other facility.

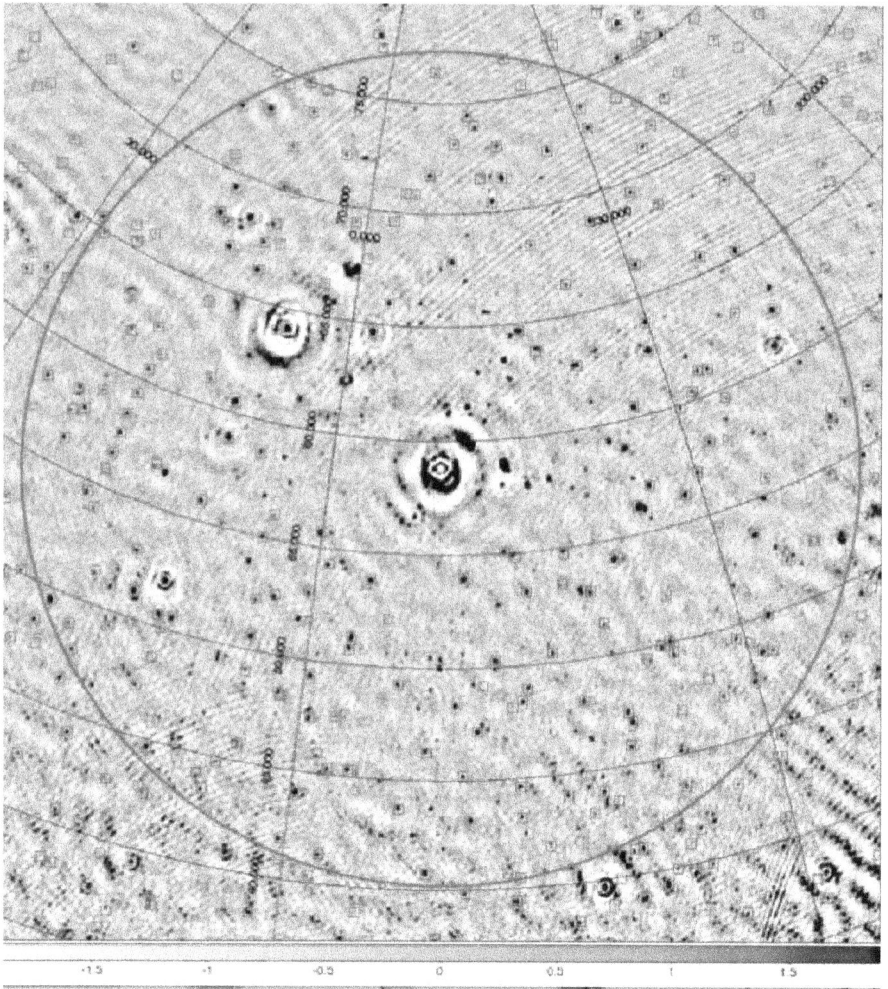

Supernova Remnant Cassiopeia A using the LOFAR array

Future Radio Telescopes

The ability to observe detailed aspects of high energy physics, analyze molecular distributions in our galaxy and catalog some of the most fascinating

111

attributes of the universe is in the realm of radio astronomy. To that end, newer and more capable instruments are being designed and are coming on line. The requirements are better resolving power, more sensitivity, and broader bandwidth.

The Atacama array of millimeter and sub-millimeter radio telescopes which is currently being assembled will cover the sky at these very high frequencies in minute detail. This will include the comprehensive mapping of many important spectroscopic lines.

An example of a new telescope design with wide band phased arrays operating in the GHz range and a lower frequency array in the background.

Conclusion

Radio Astronomy is the science of receiving signals from the Cosmos. The use of this technique has allowed the discovery of radio emissions from the Milky Way, Jupiter, Pulsars, Quasars, Supernovae, Interstellar Chemicals, Supra Luminal Jets and Planets.

Radio Telescopes have the sensitivity and resolution necessary to locate planets and environments conducive for biological development.

Chapter 4 - Spectroscopic Radio Astronomy

"The surface of the Earth is the shore of the cosmic ocean.
From it we have learned most of what we know.
Recently, we have waded a little out to sea, enough to dampen
our toes or, at most, wet our ankles.
The water seems inviting. The ocean calls."

— *Dr. Carl Sagan*

Introduction

Once the discovery of the Hydrogen line at 21 cm (1420 MHz) occurred, the rush was one for other elements and chemicals. The hydrogen line, 21 centimeter line or HI line refers to the electromagnetic radiation spectral line that is created by a change in the energy state of neutral hydrogen atoms. The change in energy state is created by the spin of the proton relative to the neutron changing its axis from clockwise to counter clockwise. As a result photons are emitted in the form of radio signals at this frequency. It did not take long for Carbon Monoxide, Oxygen, Carbon Dioxide and a host of others to be discovered. Today hundreds of elements and compounds have been detected in space and more no doubt will be discovered with the advent of more sophisticated millimeter and sub-millimeter telescopes like the Atacama array in Northern Chile.

The technique is actually quite simple with laboratory tests revealing the transition lines of the elements and compounds, frequencies verified and then a search in the interstellar regions is performed by one of many millimeter telescopes. The astronomer looks for a particular transition of a particular chemical and "dials" in the frequency. The telescope looks for and maps the emissions.

In this way maps of interstellar Hydrogen, or any other available chemical can be produced. The results of mapping overlapping chemicals tell the story of a stellar nursery, high energy phenomena or other interesting science.

One of the telescopes that brought many of these chemical

discoveries is located on Kitt Peak, outside of Tucson, Arizona.

The NRAO 12 meter millimeter telescope on Kitt Peak

This telescope has several receiver systems that work roughly in the 80 to 120 GHz and 180 to 250 GHz regions. Frequencies are examined in the same fashion as a spectrum analyzer, where a wide band portion of the radio spectrum is observed then broken down into smaller details. These regions

116

contain a multitude of chemical lines as shown below in the following plot:

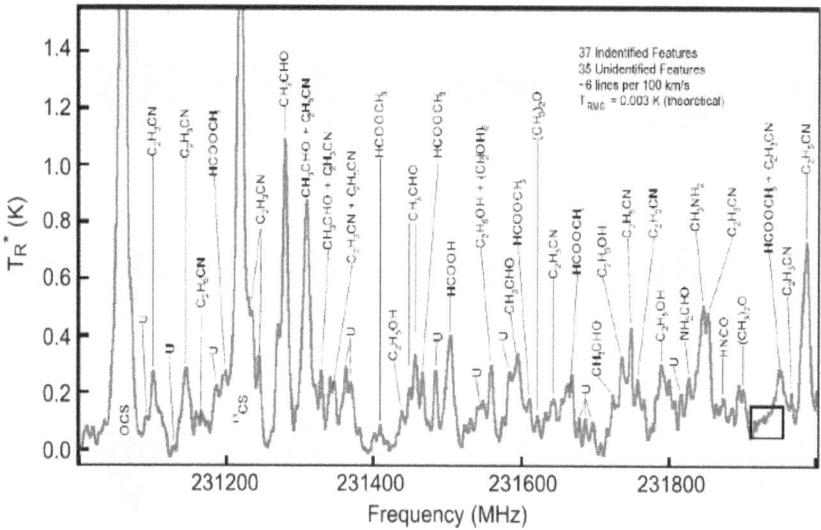

This plot show not only known chemical compounds but a significant amount of unknown chemicals. The data was taken from a millimeter telescope looking out in the cosmos. It clearly demonstrates the wealth of chemical information to be mapped as well as discovered.

The ability to look at the distribution of a particular chemical known to be associated with the process of life allows a researcher to look for those signs in space. Newly discovered planets (as of now over 2,000) will no doubt be explored with this spectroscopic technique to find these chemical signatures. This will be one of the most important steps in the ultimate discovery of life in the Universe. Once this discovery is made, the signatures will be watched for signs of change, as for instance during seasonal changes on an extra-solar planet.

The magnitude of these signatures, based on the significant size of the planet of interest, will be very high and thus detectable with current instrumentation. Patience might be required as this process of accumulating data over a long period (also known as integration) could take days months or years. Reviewing the radio astronomy formulas presented earlier will show the effects of long integration periods. Fundamentally, the sensitivity of the measurement improves significantly when long integration periods are an option.

The chart below demonstrates how the Milky Way looks at different frequencies. The radio continuum is based mostly on heat. Notice how the images change using different portions of the electromagnetic spectrum.

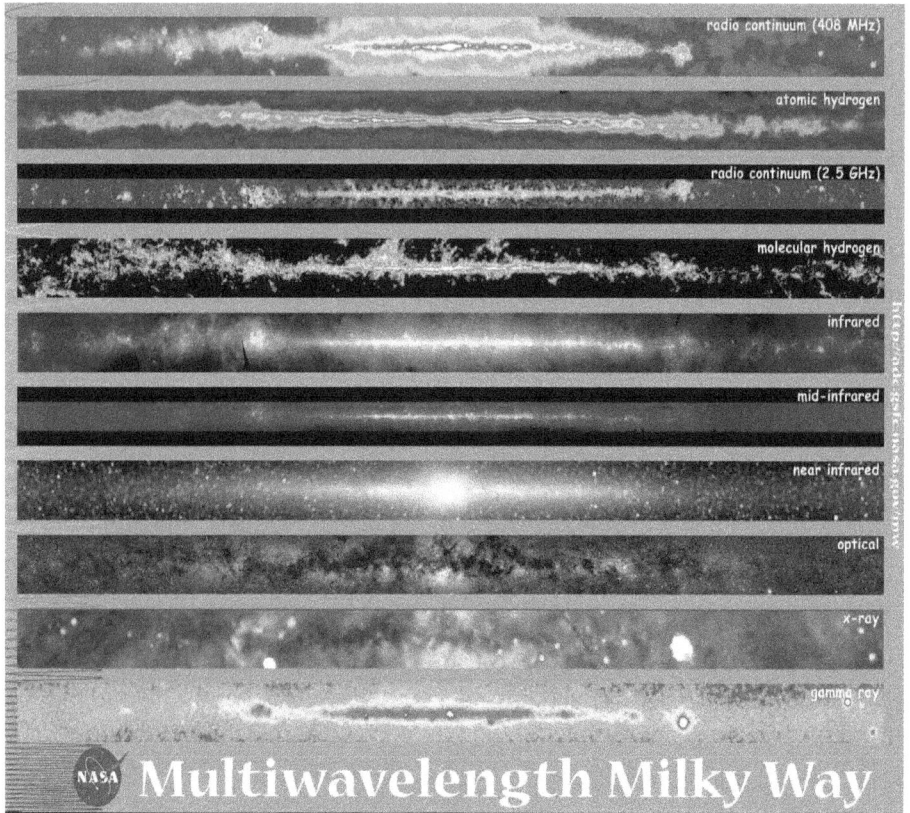

NASA **Multiwavelength Milky Way**

Much like the FM radio in your car, astronomers can tune into a particular station and by listening carefully, understand the nuances of that chemical and its relations to its surroundings.

Another unique attribute of spectroscopic radio astronomy is that the emission line positions in frequency are influenced by their motion. This is the Doppler effect, where for instance a cloud of hydrogen moving towards us exhibits a higher frequency of emission. In this way motions of molecules in the Universe can be determined.

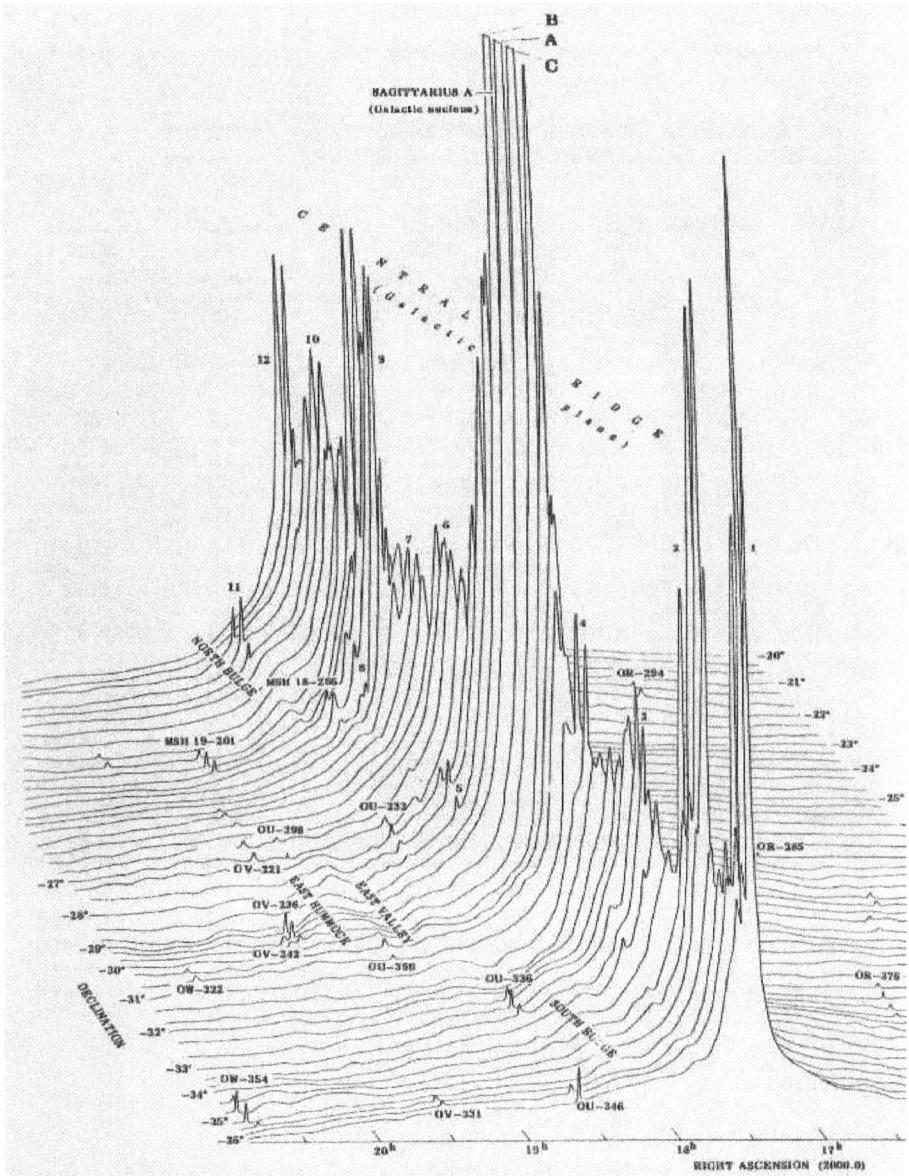

Scan of the Center of the Milky Way at 21 cm or 1,420 MHz, the Hydrogen Line

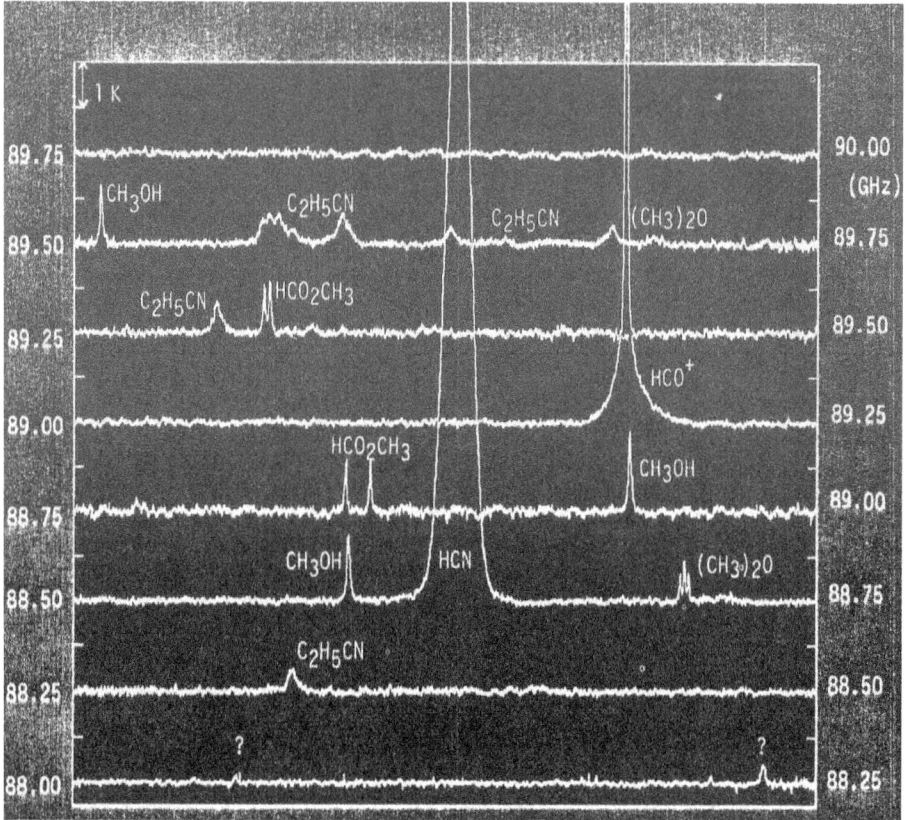

A spectral scan through Orion's Nebula showing a wealth of chemicals.

Conclusion

Spectroscopic Radio Telescopes are capable of detecting chemical signatures of life on extra-terrestrial planets.

121

Chapter 5 – Decametric Radiation from Jupiter

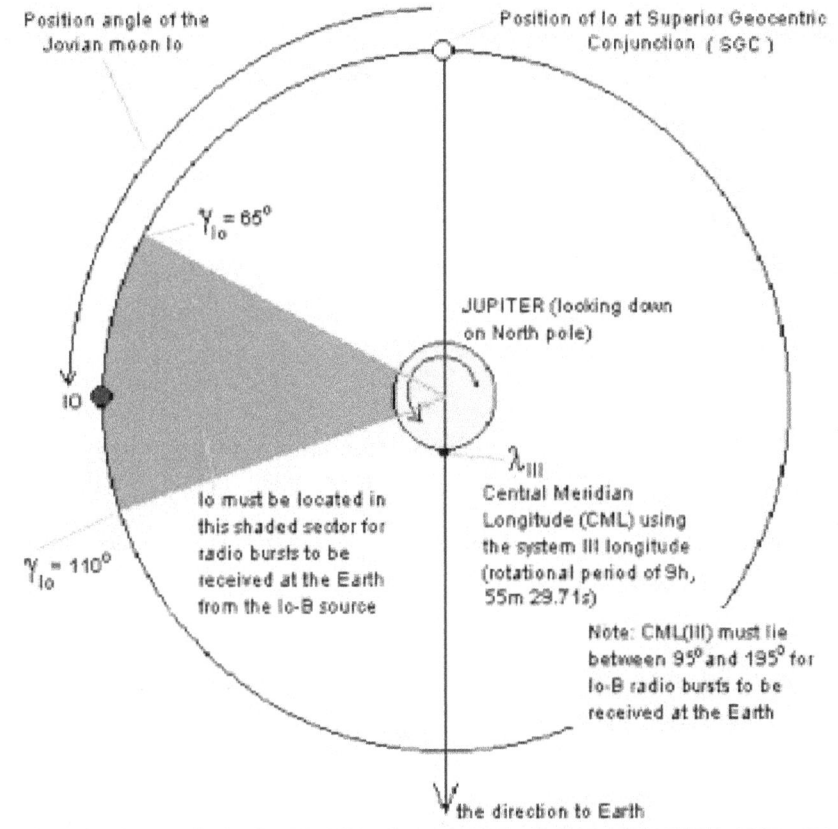

A schematic of the Jovian Radio Emissions

One of the more important and unsung discoveries of the 20th Century was the immensely strong radio emissions from Jupiter. Accidentally found in 1955 by Burke and Franklin, the ultimate measurement of energy from this phenomenon approximates 600 Billion Watts. This energy is in the form of a beam emanating from a cone and passes the Earth four times every 9 hours 55 minutes and 27 seconds.

Jupiter puts out radio emissions from the KiloHertz regions of the electromagnetic spectrum to above 10 GHz. The vast majority of energy is in the lower bands with a cutoff frequency of 39.5 MHz. This energy is easily received by standard short wave receivers using low gain antennas. It is possible that Karl Jansky in the very first measurements of the radio Universe, detected the signals, but unfortunately, all of his original data recordings have been destroyed.

Today, several simple prediction programs are available as well as receiver kits that can easily be set up by amateurs and students to observe these radio noise bursts. The Pioneer and Voyager series satellites all had swept frequency receivers to measure and map this radiation as they flew to and around Jupiter. With this data, a model of the generating mechanism was developed. Fundamentally, it is generated by the magnetosphere of the planet which concentrated high energy electrons in its field and allow radiation to be produced in a cone like shape around the northern and southern magnetic poles. Io, the closest moon of Jupiter influences these emissions in a predictable way.

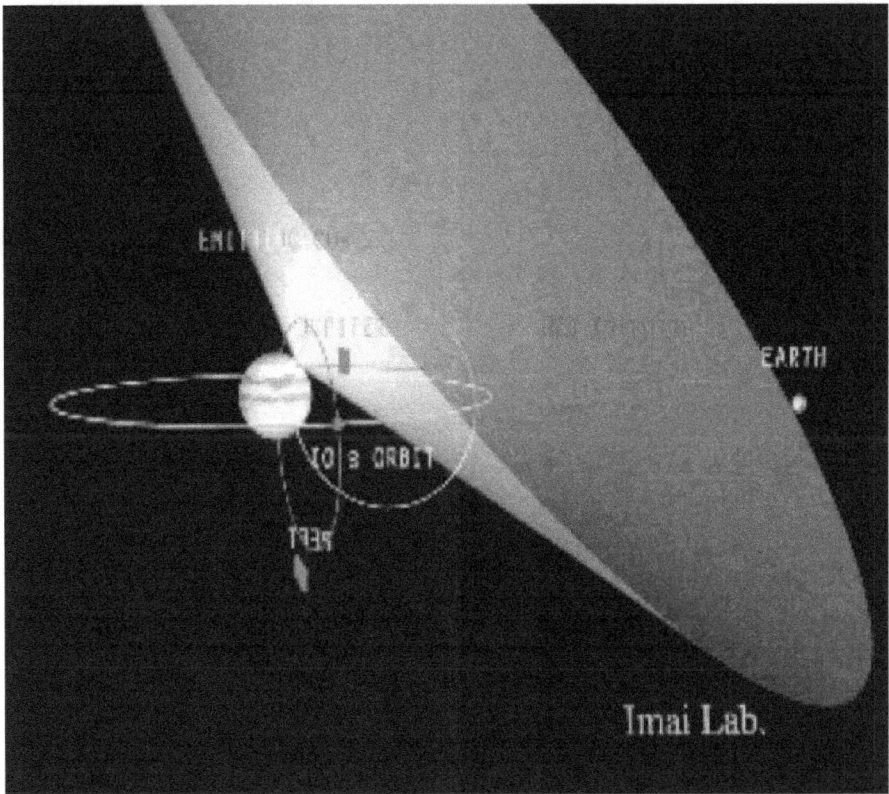

Edge on view of the radiating cone from Jupiter's Magnetosphere

When electrons and protons move through a magnetic field their paths are changed, the particles are accelerated and move in spirals around the magnetic field lines towards either the north or south pole. These charged particles emit radiation based on the amount of acceleration they experience. This radiation in the case of Jupiter is in the radio frequencies and increases with an increase in magnetic field strength. The specific type of radiation is called cyclotron emission after a type of particle accelerator. The radio emissions are not heard above 40 MHz as this is dependent on the strength of the magnetic field, the magnitude of this field can be estimated based on these emissions. [1]

The enormous power broadcast by this planet greatly exceeds any other planets' and due to its very regular movement, acts like a slowly rotating pulsar. No other radio source, especially those from our own Earth, come close to this power intensity in our solar system.

Considering the power level of hundreds of billions of watts, it is conceivable that extrasolar planets could be capable of detecting it. 600 billion watts is equivalent to 147.8 dBm (a logarithmic depiction of power). As we will see in a later chapter, when we consider how much power is lost over interstellar distances and combine these numbers, good receivers and antenna systems have a chance to detect the Jovian like emissions from other solar systems. It therefore would be prudent to launch a search for Jupiter like planets in other star systems.

The importance of Jupiter is just being realized as one of it main benefits is that it sweeps up many meteorites and comets, some of which could be a danger to the Earth. There are some scientists who feel that a Jupiter like planet is necessary for survival in a solar system. If this is so, then the probability of life in other star systems increases.

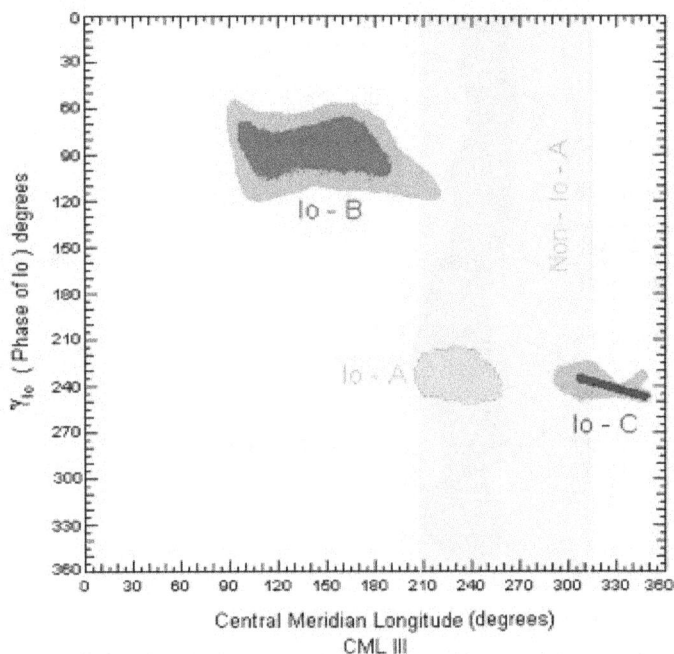

The location of Radio emissions from Jupiter as a function of Longitude on the planet

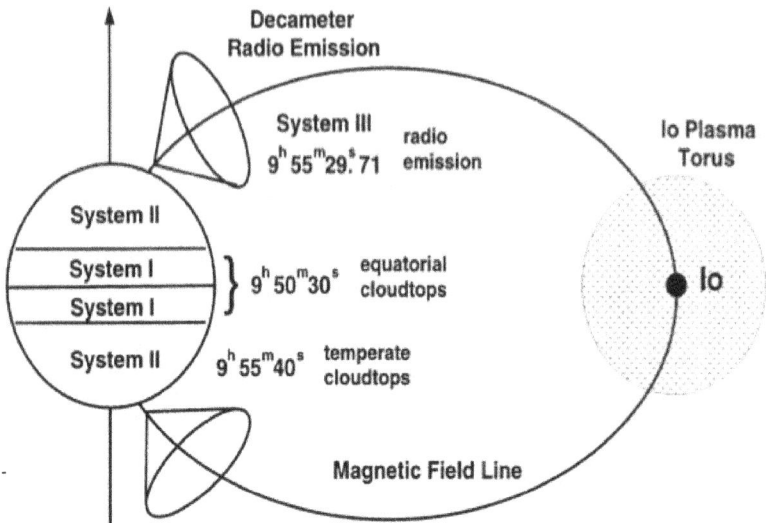

Another schematic of the Jupiter Radio Emissions

The Jupiter radio emissions span a band of frequencies mostly from 1 to 39.5 MHz and have had several features identified as separate entities. There are long and short bursts that are based on separate physical phenomena. The short bursts can be a little as milliseconds long whereas the long bursts can take many minutes. The amount of integrated energy over the whole period when the phenomenon is observed is substantial. This is important as any planet trying to observe these emanations can do so by looking at very slow changes in the cosmic background at these low frequencies. Another way of saying this is that the modulation of the emanations are within a very small bandwidth, a distinct advantage in detecting low level signals. Listening to these emissions with headphones or a speaker reminds one of whistles and pops, changing in tone and tenor. A graph of the spectrum reveals energy transferring from one frequency to the next in a sometimes picturesque looking display like the chart below.

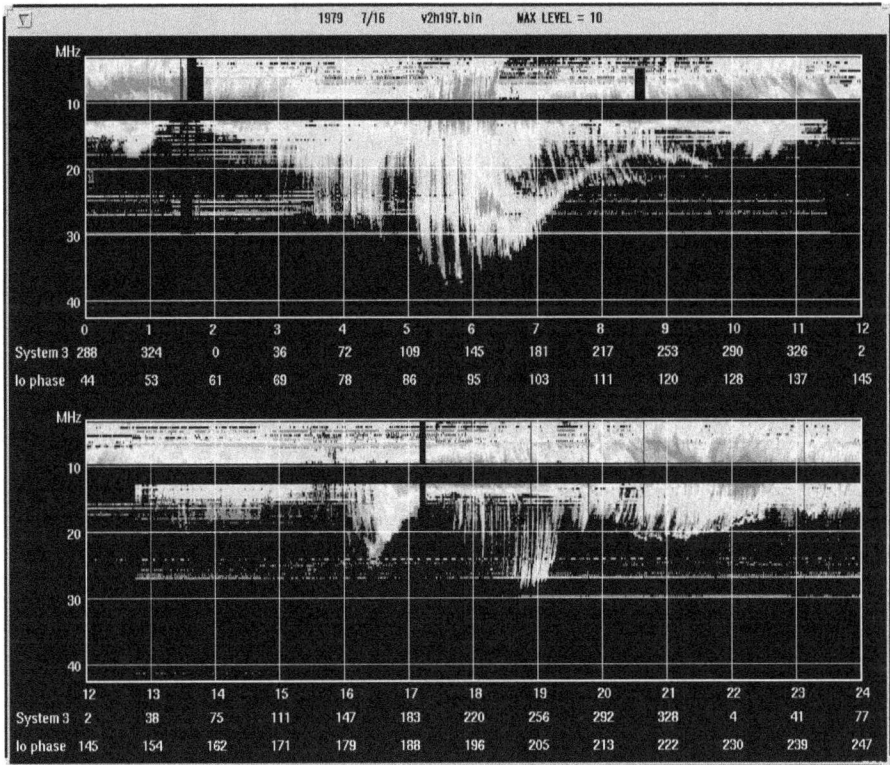

1979 7/16   v2h197.bin   MAX LEVEL = 10

	0	1	2	3	4	5	6	7	8	9	10	11	12
System 3	288	324	0	36	72	109	145	181	217	253	290	326	2
Io phase	44	53	61	69	78	86	95	103	111	120	128	137	145

	12	13	14	15	16	17	18	19	20	21	22	23	24
System 3	2	38	75	111	147	183	220	256	292	328	4	41	77
Io phase	145	154	162	171	179	188	196	205	213	222	230	239	247

The elegant songs of Jupiter

NASA has created the *Radio Jove* project to interest students and young radio astronomers in the Jupiter radio outbursts. This project contains software that predicts the time of the outbursts and directs the viewers to inexpensive equipment that can be used to receive and record the emissions. The antennas and receiver in this system work in the 20 to 25 MHz shortwave bands. This particular frequency span takes advantage of lower ionospheric attenuation and is closer to the peak of the emissions.

Closer examination reveals that there are polarization characteristics that give hints as to how the emissions are created. For the most part, the lower frequency outbursts are randomly polarized, whereas the higher frequency outputs are right or left hand circular.

Research still is being done to better understand how this immense beam of energy is created. But for the purposes of this book the important issue is that it is incredibly high power, directional and very predictable. It seems probable that other gas giants recently discovered orbiting around other stars will have similar broadcast methods (cyclotron radiation) and hence be detectable with careful measurements here on Earth.

Neptune also emits a much smaller amount of radiation between .5 and 2 MHz. The Earth emits an even smaller amount of radiation below about 2 MHz. Fluctuations can be observed as the solar wind moves the position of the magnetosphere. The Sun itself is a powerful emitter which has a steady component based on its black body radiation and a variable component based on solar flare activity. Uranus and Saturn also emit radio radiation, but magnitudes below that of Jupiter.

An interesting feature of the pattern of energy around Jupiter as measured at shortwave frequencies is that is looks very similar to an antenna pattern, with a main beam, sidelobes and a front-to-back ratio.

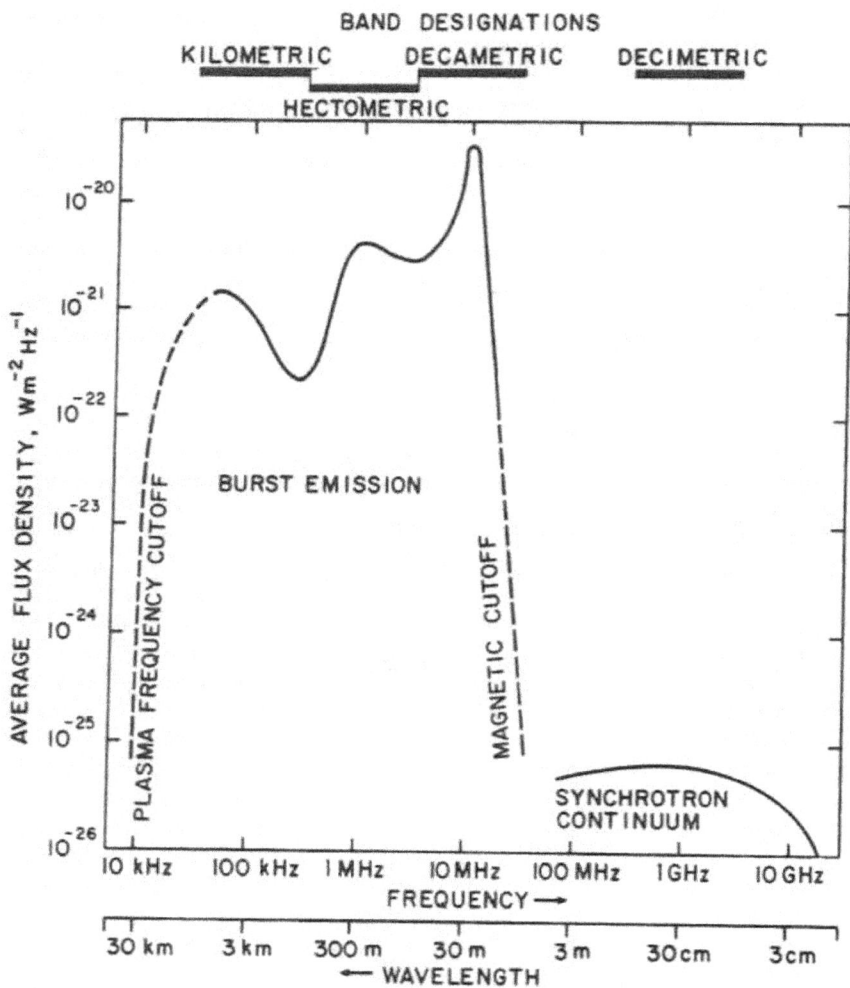

The Spectrum of Jupiter Radio Emissions

Conclusion

Jupiter and Jupiter like planets can emit significant amounts of radio transmissions. Jupiter itself radiates over 600 Billion watts in a very predictable way and approximates a slowly rotating pulsar.

Many other recently discovered extra-terrestrial planets are Jupiter like. If they share the same magnetic field structure as Jupiter, then it is likely that they emit significant amounts of power in the lower ranges of the radio spectrum.

Reception of Jupiter like emissions from extra-terrestrial solar systems could be the indication that these systems contain Jupiter like planets that serve the same benefits to their solar neighborhood as our gas giants do. This includes sweeping the local space of asteroids, comets and other debris that could potentially crash into life supporting planets.

Chapter 6 - The Search for Life with Radio Telescopes

Introduction

An ultimate consequence of thinking about life on other planets is an attempt to find it when the equipment seems capable. This is exactly what happened beginning in 1957 when project OZMA (detailed later) was initiated. Several other attempts have been made since, some using antennas as big as 1000' in diameter. To date nothing has been found. Arrays of antennas have been designed specifically for this search, costing many millions of dollars. They have been unsuccessful as well. The search for interstellar life in these cases is based on other intelligent civilizations emitting modulated energy which will be observable from Earth.

There are other methods using the same instruments that look for other types of evidence of life. In these cases, spectroscopic methods are used to isolate specific chemicals that indicate the process of life on a planet. This approach covers many more forms of life as the process of photosynthesis as well as animal activity produces these signatures.

The chronology of these efforts was described by Guiseppe Cocconi and Philip Morrison suggesting in the journal *Nature* in 1959 that microwave interstellar communications is possible.

In 1963 Russian scientist Dr. Kardashev examined quasar CTA-102, the first Soviet effort in the search for extraterrestrial intelligence (SETI). In this work he came up with the idea that some galactic civilizations would be perhaps millions or billions of years ahead of us, and created the Kardashev classification scheme to rank such civilizations. Other notable experts in the USSR were Vsevolod Troitskii and I. S. Shklovskii who was Kardashev's professor. [1]

Project OZMA

In 1957, a scientist working at the National Radio Astronomy Observatory began a project to examine the possible radio transmissions from several well chosen stars. In 1960 Frank Drake used the 85' Howard Tatel telescope centered at a frequency of 1,420 MHz (the Hydrogen Line) and examined the spectrum above and below the center emission frequency in hopes that any intelligent life capable of transmission would recognize the importance of this Line and use it as communications channel to others who might be listening. Drake used standard high quality short wave receivers to scan around the Hydrogen line, recording his results for later examination.

The stars Tau Ceti and Epsilon Eridani were scanned and the spectrum scanned over 400 KHz with a bandwidth of 100 Hz was recorded for over 150 hours without positive results.

This is perhaps the first professional attempt to look for life elsewhere in the Universe and revealed the serious interest both professionally and non-professionally of the possibility of other intelligent life forms attempting to communicate. Although the stars examined are relatively close, the amount of energy required to produce a believable result in this experiment would have had to be extraordinarily high.

A second experiment, called Ozma II, was conducted at the same observatory by Benjamin Zuckerman and Patrick Palmer, who intermittently monitored more than 650 nearby stars for about four years (1973–76), again without any discernible signals

85' Howard Tatel Radio Telescope used by Frank Drake on Project OZMA

134

Project Cyclops

During the 1960s, NASA scientists first performed more realistic calculations with regards to the required sensitivity for a radio telescope capable of finding radio transmissions from other worlds. This project was extremely ambitions and showed the magnitude of the required effort to detect extremely faint signals, like those that would be transmitted by extra terrestrial intelligence.

The telescope was actually composed of one thousand 100 meter antennas and was to be placed in a radio quiet area in New Mexico. Each antenna would track the same radio source and send its information to a central building which would combine the data with all of the other antennas to produce a very sensitive recording of the area of interest in the sky.

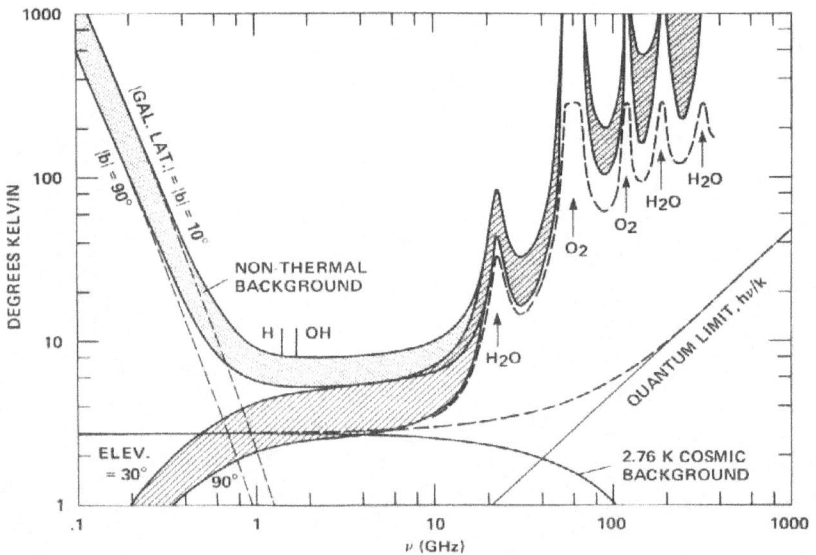

Radio Window from 1 to 10 GHz, sometimes referred to as the "water hole"

The frequency of operation was to be 1-10 GHz, which is now known as the "radio window" (or water hole) as it has the best transmissivity from space to the ground. Below 1 GHz, terrestrial noise can compromise data and above 10 GHz atmospheric absorption can attenuate signals.

Cyclops close up view of 100 m antennas and control station

The Cyclops Report

"Absence of evidence is not evidence of absence."
-Martin Rees

A comprehensive report on the justification for the building of the Cyclops Array was written in 1971 during a summer faculty fellowship program

136

at Stanford and Ames Research Center. The report presents a cogent, well thought out approach to the design and use of a very large telescope designed for radio astronomy research as well as the search for extraterrestrial intelligence. Excerpts follow:

"The picture we behold is one of *cosmic evolution:* a universe that was born has evolved into galaxies of stars with planets, and into living things, and that will continue to evolve for aeons before vanishing into oblivion or collapsing toward rebirth. We owe our very existence to strange and wonderful processes that are an inherent part of this cosmic evolution. For example we know that heavy elements were not present in the beginning but are formed in the fiery interiors of stars, many of which die in a titanic explosion. The dust of the heavy elements hurled into space becomes the raw material from which new stars with earth-like planets can form. The calcium in our very bones was manufactured
in the cores of countless long dead stars not unlike those that sparkle in our night sky. The ability of intelligent life to contemplate its own existence has always fascinated the philosopher. Now we see the even more fascinating picture of the entire universe contemplating itself through the minds and eyes of the living beings evolved by the universe itself.

Out of the findings of the sciences that have converged to give us our present picture of cosmic evolution we draw certain conclusions that are the premises of a plausibility argument for the prevalence of technologically advanced extraterrestrial life.

1. Planetary systems are the rule rather than the exception. Our

137

understanding of   formation leads us to expect planetary systems around most stars, and to expect   a favorably situated planet in most systems. There are therefore probably on the   order of 10^{10} potential life sites in the Galaxy.

2. The origin and early evolution of life on Earth is apparently explicable in terms of the basic laws of physics and chemistry operating in the primitive terrestrial environment.

3. The laws of physics and chemistry apply throughout the universe. Moreover, the composition of the primordial material, from which life on Earth arose in accordance with these laws, is commonly repeated elsewhere.

4. The factors causing natural selection, which led to the evolution of more complex species and ultimately to intelligent life on Earth, may reasonably be expected to be   present on any planet on which life originated.

5. Intelligence, if it develops, is generally believed to confer survival value and   therefore to be favored in natural selection. On Earth, intelligence led to attempts to   modify and utilize the environment, which were the beginnings of technology. This   course of development seems a highly probable one on any earth like planet."

Distant View of 1000 100m antennas

Ultimately the Cyclops array was never built, but the premise under which is was proposed is still relevant today.

High Resolution Microwave Survey (JPL)

In 1975 NASA began funding definition studies for the Search for Extraterrestrial Intelligence (SETI) program. After working at low funding levels for several years NASA began was to have been a $100 Million formal SETI effort that had been renamed the High Resolution Microwave Survey (HRMS). This was initiated in 1992 and terminated in 1993 after Congress

139

cancelled the funding.

The system consisted of a series of receiver systems that were to be attached to existing Deep Space Network (DSN) antennas operated by NASA. The frequency of operation was in the 1 – 10 GHz range, where minimal attenuation effects from atmospheric water exist. Feed horns were designed using wideband corrugated approaches followed by dual polarization ortho mode transducers. This gave the system the ability to receive all types (linear and circular) polarizations. Following the feed horn was a series of low noise amplifiers that then feed down converters for eventual transmission of the received signals to spectrum analyzers and other data acquisition hardware. The final piece of hardware developed was know as the Multichannel Spectrum Analyzer which was capable of analyzing 14 Million 1 Hz channels.

The HRMS system was operational starting in October 1992 at the 1000' radio telescope located in Arecibo, Puerto Rico. The antenna was initially pointed at Gliese 615.1A and measurements were made for about a year. A second system was used at the Goldstone facilities of the Deep Space Network (DSN) for a period ending in 1993, when the funding was cut. The project won the "Golden Fleece Award" form senator William Proxmire as a waste of taxpayer's money. [4]

The NASA SETI Program Patch

Project Phoenix

The SETI Institute, a privately funded effort, started project Phoenix which ran from 1995 or roughly after the end of HRMS to 2004. The core science and engineering team from HRMS was retained and with their efforts rebuilt and upgraded the existing equipment. They surveyed more than 700 sun like stars within 200 light years of Earth in the frequency range of 1.2 to 3 GHz looking for narrowband signals indicative of extraterrestrial communications. Telescopes in Australia were also used with this equipment as well as the 43 meter Green Bank radio telescope.

SETI@Home

This is an internet based public volunteer distributed computing project that uses data taken "piggy back" from the Arecibo radio telescope and analyzed on personal computers. The format is either a screen saver or a program that works in the background. The software searches for five types of signals that distinguish them from noise:

1. Spikes in the power spectra
2. Gaussian rises and falls in power
3. Triplets – three power spikes in a row
4. Pulses that possibly represent narrowband digital transmissions
5. Signal waveforms detected by autocorrelation

With over 5.2 million participants worldwide, the project is the largest distributed computing project with the most participants to date. The original intent of SETI@home was to utilize 50,000-100,000 home computers. Since its launch on May 17, 1999, the project has logged over two million years of aggregate computing time. On September 26, 2001, SETI@home had performed a total of 10^{21} floating point operations. It is acknowledged by the *Guinness World Records* as the largest computation in history. With over 278,832 active computers in the system (2.4 million total) in 234 countries, as of November 14, 2009, SETI@home has the ability to compute over 769 TeraFlops. For comparison, the K computer, which as of June 20th 2011 was the world's fastest supercomputer has achieved 8162 TeraFlops. [1]

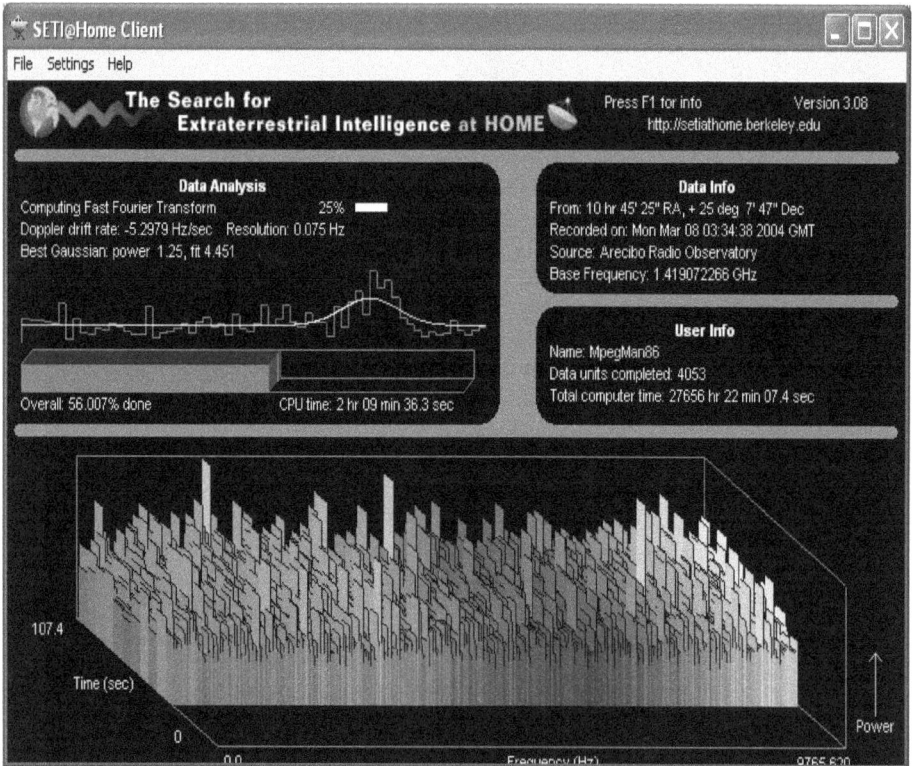

SETI@Home Operations Screen

To date, no convincing signals have been found, however it has demonstrated the power of distributed computers and the magnitude of the public interest in SETI.

The Allen Array

This is an array of 42 parabolic antennas placed in northern California and used for the primary purpose of SETI searches. The operational frequencies are between 0.5 and 11.2 GHz, again in the "water hole." It received significant funding from Paul Allen, co-founder of Microsoft, hence the name. The ultimate configuration is for 350 antennas but funding currently will not support this effort.

The SETI Institute manages this project and currently employs some 50 researchers for SETI searches as well as other astrophysical explorations.

One of the 6 meter Allen Array Elements

1000' Radio Telescope and Radar in Arecibo, Puerto Rico, used for several SETI searches

Amateur Efforts

Due to high interest in the public, several amateur radio astronomy groups have formed and have commenced SETI research. These groups are made up of radio amateurs, technicians, engineers and scientists who have refurbished several large defunct antennas, including a 140' dish in California and several 60' dishes in Colorado. A short list of these groups follows:

> SETI League
> Bambi
> Deep Space Exploration Society

All of the groups mentioned have web sites.

Conclusion

Significant and expensive efforts have been employed in the past to look for intelligent transmissions from space. For the most part, these efforts use existing radio telescopes normally used for scientific research on physical processes in the cosmos. To date, no signals that indicate sentient life have been found.

The Cyclops Report was one of the most articulate, well thought out studies to define what would be required to give the best chances of detecting extra-terrestrial signs of intelligence.

"Piggy Back" receivers have been used to minimize interference with ongoing scientific research on radio telescopes.

There has been a significant amount of public interest in participating in extra-terrestrial signal detection.

Chapter 7 - Anatomy of a Received Signal

The 18 Meter Prime Focus Antenna

During the period of November 1, 1993 to November 10, 1993, signals around 430 MHz were received with the above picture antenna from a sidereal source. The National Office of Telecommunications / Institute of Telecommunicational Sciences owns this 18 meter dish and during this period, allowed research to be conducted by outside scientists and engineers. One organization, the Deep Space Exploration Society (DSES) has refurbished the antenna, upgraded the motion control and had purchased Hydrogen Line spectrometers to look at the Milky Way.

After several occurrences of the radio source it was facetiously labeled Little Green Men 1 or LGM1.

In the months preceding the detection of LGM1 the author had designed and built a feed antenna, low noise amplifier and frequency down converter to map the radio universe at 430 MHz from outside Boulder, Colorado. This mapping exercise was conducted for several months and the area around zenith was scanned at +/- 20 degrees in one degree increments. As the Earth rotates, the antenna beam correspondingly scanned across the sky, along a course of constant Declination. At an elevation of 90 degrees, a strong signal was detected at the location of +40 08' 54" degrees declination, 14h 06m 25s of Right Ascension, just outside the constellation of Ursa Major. The general position is off the "handle" of the Big Dipper, with a distance the same as the separation between the first two stars as shown in the star map below:

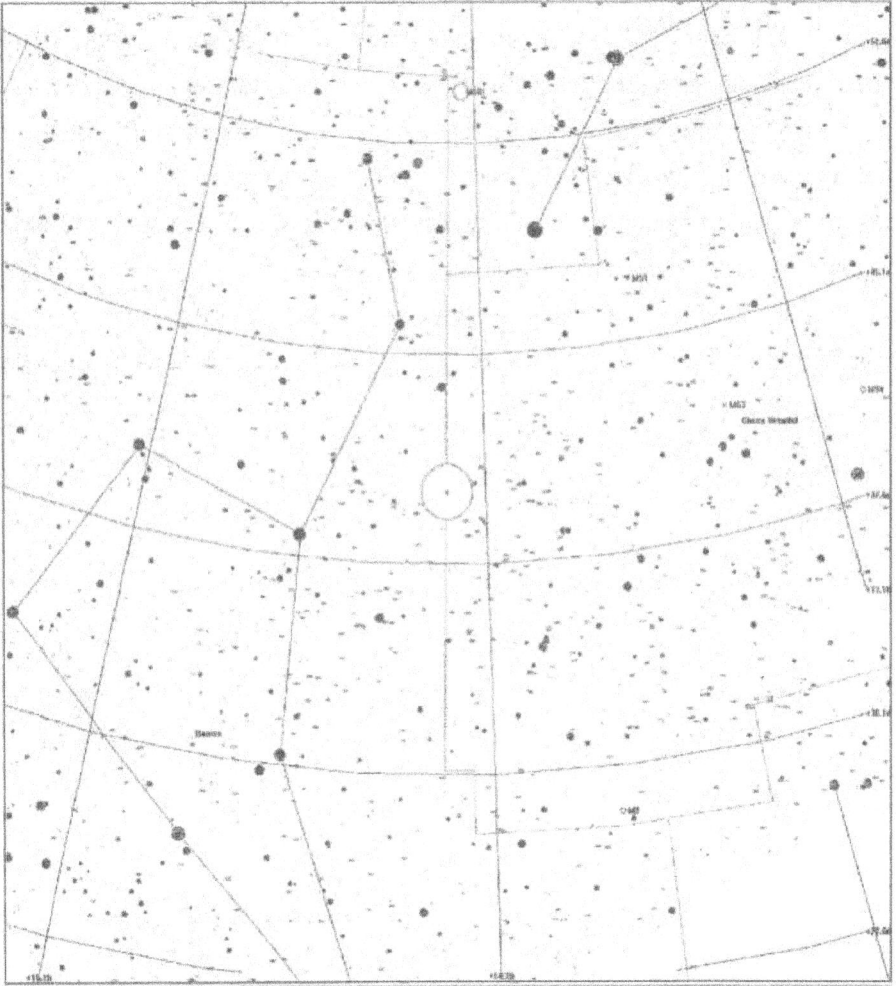

Sky Map of the position of "LGM1," the upper two stars on the right side are the "ladle" stars of the Big Dipper. The circle shows the position of the radio source as well as the diameter of the antenna beam.

This particular signal rose up several dB above the noise floor which calculated to over 7000 flux units (Jansky's) of power per square meter. The gain of the antenna is about 35 dBi at this frequency with a beam width of 2.4 degrees. The Low Noise Amplifier (LNA) had a noise temperature of 0.8 dB and a gain of 25 dB. The equipment that followed, sometimes referred to as the "back-end" consisted of a high quality short wave receiver set at 13 KHz bandwidth. The audio output was sent to a diode detector and integrator circuit, then onto a 10 bit digitizer that was connected to a computer that recorded the results. The data was reduced by a Savitsky-Golay routine to present the following plot:

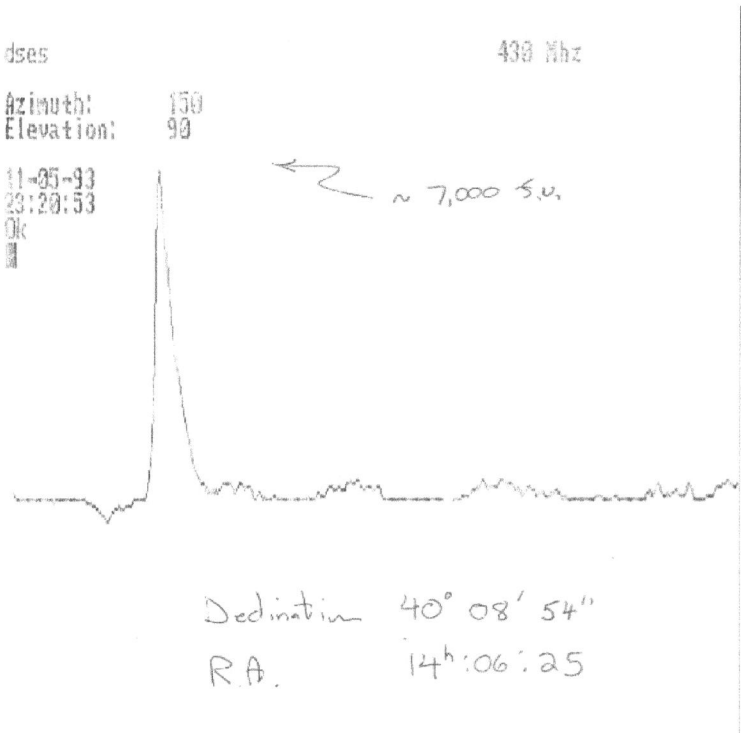

The original plot of "LGM1" from the smoothed data.

The signal emerged from the noise at the beginning of the observation period, rose in amplitude over the next several days and receded into the noise after about a 10 day period. At its peak, the signal level was strong enough to observe the antenna's sidelobes. Audio recordings were made which sounded like horses galloping over a cobble stone road.

Significant excitement was generated from this signal but it was clearly recognized that a thorough investigation was in order and so the philosophy was adopted that it was not of extraterrestrial origin and therefore the quest was focussed on determining the true origin of the signal. Put in other words, to assume that signals of this nature are man made or at the very least easily explainable by geo-magnetic activity or other process, is appropriate for serious consideration. All options must be considered and the researcher must pass every one of the "prove me wrong" tests. If the researcher publishes his or her work before completing a rigorous examination of the facts, reputations of an otherwise competent researcher can be damaged beyond repair.

The signal received was measured on several occasions and the following list of qualifying exercises was performed:

The Cheyenne Mountain facility outside Colorado Springs, Colorado was contacted and asked about the existence of satellites that could have been transmitting at the same frequency and in the same location at the same time the observations were made. After research into the data files stored at the facility and a review of satellites that transmitted on or near the frequency of interest, no match was found.

The National Radio Astronomy Observatory was contacted and a detailed report was sent to people at the Very Large Array (VLA) and Very Long Baseline Array (VLBA), both located in Socorro, New Mexico. Initially interest in pursing this signal was not deemed important enough to displace scientists who in many cases waited years to use these facilities. A few days after the initial contact, NRAO recontacted the author and had a subsequent conference call with people from the Westerbork Synthesis Array in the Netherlands, a world class radio telescope facility. The reason for the call was based on the fact that Westerbork had observed signals at this frequency at the same time and place in a sidelobe of their array during a standard observation run. All details of the 18 meter dish recordings were discussed but nothing conclusive was determined. NRAO used a portion of the VLA and a portion of the VLBI to observe radio emissions in the same area of the sky. These instruments do not have the capability to observe 430 MHz and as a result no confirming signals was detected.

The Search for Extra Terrestrial Intelligence (SETI) group was contacted but did not offer an opinion.

Local media was contacted and invited to come in one morning to record the signal at the predicted time of transit. They arrived to observe the signal at exactly the predicted time. The DSES group was careful to keep the source defined as unknown until further research was completed.

After more spectral analysis of the signal, it was decided that a modulated portion had been detected, probably from an amateur radio operator within the passband of the receiver. Local interference was thus a cause of some of the audio recordings. When looking within the envelope of the demodulated signal it was noticed that it contained sin wave bursts with a phase reversals. Spread Spectrum modulation techniques, especially direct sequence, are characterized by this type of modulation.

Conclusion

Many signals are intercepted by radio telescopes and other receiving apparatus. Most if not all are explainable by carefully examining the data, the environment, and the equipment. As there have been several cases of ephemeral, high intensity signals emanating from the radio sky at various frequencies, it is generally believed that these are attributable to extra-stellar phenomena like novas and supernovas. Other explanations are possible, but only with further research and phased array designs like the Shoemarkian array, which can detect radio emissions from extensive portions of the sky simultaneously.

It is also imperative to take a purely scientific approach to determining where signals of this type are from as it could easily be generated by a satellite or local interference.

Chapter 8 - Interstellar Communications Fundamentals

Introduction

Electromagnetic waves dissipate energy at the rate of $1/R^2$, where R equals the distance between the source and receiver. This law is central in calculating the amount of radio energy that would be expected to be received by a radio telescope on Earth from transmitters situated in interstellar space or conversely from transmissions from Earth to those receiving our signals in space. Other important variables are the size of the antennas at both "ends", the power of the transmitter, the sensitivity of the receiver and any advantage gained from certain modulation techniques. This law is dependent on frequency as well as distance thus:

$$Free\ Space\ Path\ Loss = \left(\frac{4 * pi * d * f}{c} \right)^2$$

where:

d = distance

f = frequency

c = speed of light

pi = 3.1415926

This formula reduces to a logarithmic presentation typically used in communication theory. This is one of the first steps to understand the necessary equipment to make interstellar contact by designing the antenna.

To start any antenna design, the engineer has to understand the basics of the communications link. Fundamentally, this includes a transmitter, receiver, associated antennas and associated feed cables. Also, there is a distance between the components that dictates the amount of loss to be expected.

Knowing the characteristics of all but one of the components will give the designer the ability to calculate the missing value of the remaining one. For instance, if we know everything but the required receiver antenna gain, following the next set of formulas will show what is required to complete a successful link.

Keep in mind that the path loss is actually an ideal number, in real life designs it is important to provide a link *margin* to compensate for variations like weather and obstructions.

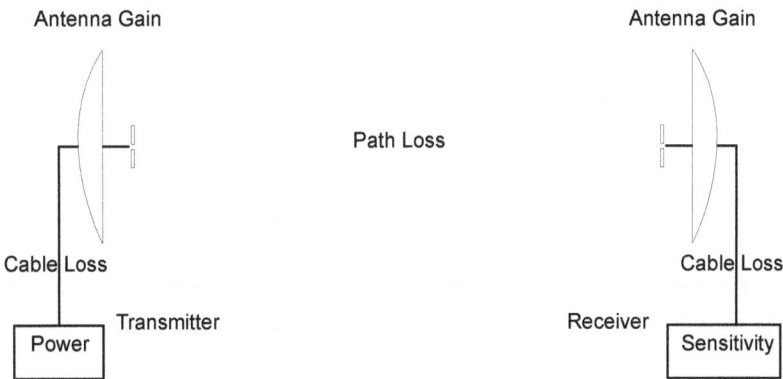

An Idealized Transmitter to Receiver Link

To calculate how much gain is needed for an antenna, one should look at both ends of the link, for instance the transmit site and a receive site. In radar systems, one antenna typically does both duties, but due to some other system details a more refined radar equation is needed to define the radar link.

To understand the link analysis for a two antenna system, several factors need to be quantified:

Distance between antennas to calculate path loss

Gain of the transmit antenna

Power of Transmitter

Sensitivity of Receiver

Gain of the receive antenna

Losses in feed cables

To calculate the space loss between antennas with existing antennas use the following formula:

$$Path\,Loss\,(dB) = C + (20 * \log(F * R) - G1 - G2)$$

where,

F = Frequency in MHz

R = Range in units defined by C

G1 = Gain in dB of antenna 1

G2 = Gain in dB of antenna 2

C =   32.45 for Kilometers

37. 80 for Nautical Miles

36.58 for Statute Miles

-27.55 for Meters

-37.87 for Feet

Calculating just Path Loss can be done by using the following equation:

$$Pathloss = C + 20 * \log(F * R)$$

where,

F = Frequency in MHz

R = Range in units defined by C

C = 32.45 for Kilometers

38. 80 for Nautical Miles

36.58 for Statute Miles

-27.55 for Meters

-37.87 for Feet

Adding in the effects of Transmitter Gain, Cable Loss, Receiver Sensitivity and any conversion gain (added gain due to efficient modulation schemes) yields:

$$Link\ Margin = Path loss - Pt - G1 + Tl - |(RTH)| - RG + RLO - CG$$

where,

P_T = Transmitter Power in dB

G1 = Transmitter Antenna Gain (dB)

Tl = Losses in Transmitter Cable (dB)

RTH = Receiver Threshold in dB

RG = Receive Antenna Gain in dB

RLO = Losses in Receiver Cable (dB)

CG = Conversion Gain in dB

Conversion gain is an important concept in advanced communication where more distance can be covered or less power used relative to conventional designs. This comes in the form of a mathematical way of modulating and de-modulating a particular signal. The important issue is that the modulation scheme needs to be known a priori as the detection of just the power of a cleverly modulated signal can be challenging. A case in point is the standard GPS signals now covering the Earth. The actual signal levels are extremely faint on the order of -145 dBm, well below the noise level of a standard receiver. How the signals are detected so reliably is by use of spread spectrum techniques. Fundamentally, the log base 10 of the ratio of the modulated signal to the spread signal gives whats known as "processing gain" which can be added to the link equation above to evaluate the quality of the communications path. Unfortunately, someone who is attempting to receive a spread spectrum has to know an encryption code to de-modulate the signals. This fact is used on purpose for military and other covert activities.

The Role of Earth's Ionosphere

Another factor in the propagation of radio signals is the ionosphere which consists of several layers of ionized particles high in our atmosphere.

Oliver Heaviside became famous for reducing Maxwell's 20 equations with 20 variable each to a much simpler set of fo ur which are used extensively today in the design of antennas and other devices. After his work, antennas, transmitters and receivers were designed starting in the lower frequency regimes and progressing upward. At the lower frequencies, great distances could be covered by modest powers and modest antennas. This is due to the Earth's reflective Ionosphere, where several layers of ionized particles form layers that change over the period of a day. These layers have been called the Heaviside Layer in honor of his achievements. The sun's 11 year cycle has an effect on these layers as the amount of charged particles from flares and the solar wind modify the reflective properties of the Ionosphere as well.

Shortwave reception improves during the evening hours and is very limited during the day. This implies that low frequency radio astronomy is best done during the daytime. The Maximum Usable Frequency (MUF) is calculated on an hourly basis from propagation and space environment centers then published on the Internet. Only above the MUF can signals propagate into or from space. This is important to remember when considering what we have sent to the cosmos. In the history of radio starting at very low frequencies, it follows that very little has made it to the stars.

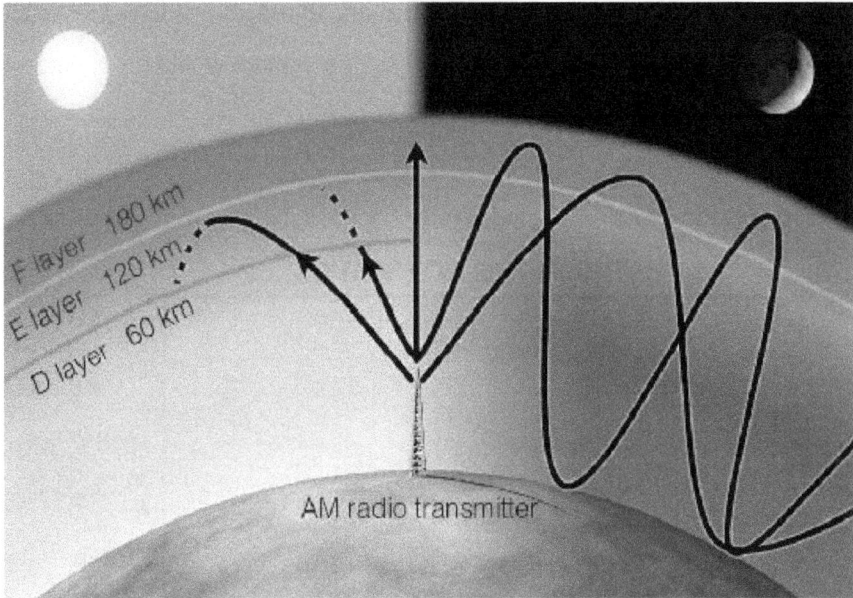

F layer 180 km
E layer 120 km
D layer 60 km
AM radio transmitter

Radio Broadcasting below 30 MHz is highly influenced by the Ionosphere during the day and night

Other factors effecting Communications

On Earth, and presumably on other similar planets, attenuation of the signals is also effected by water vapor. This is the primary reason for the "Water Window" in radio astronomy. It also effects radar transmissions and other high power operations. Rain in all forms effects propagation as well including radio, Infrared, and Visible portions of the electromagnetic spectrum.

The following graphs illustrate the magnitude of the attenuation. This clearly will effect interstellar communications. The solution of course is to transmit and receive outside of the planet's atmosphere, as in space, on lower frequencies, down to 1 GHz.   The cost of such a venture would be significant.

ATMOSPHERIC ATTENUATION

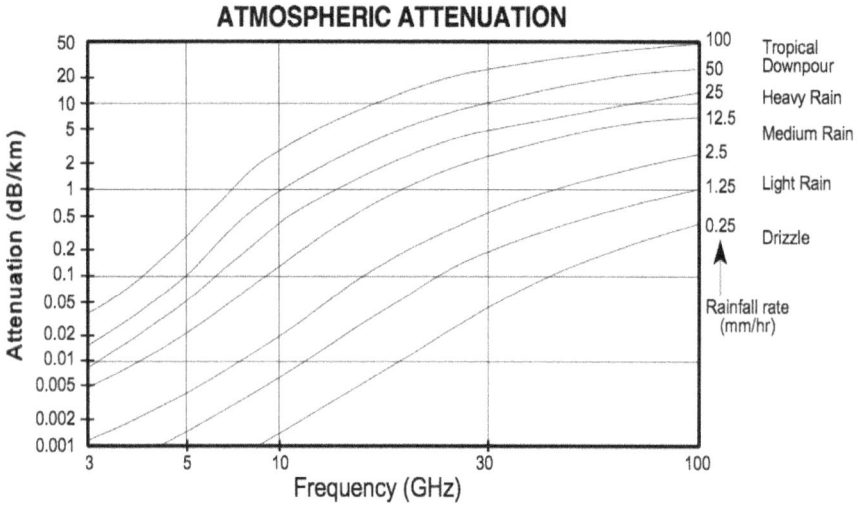

The Effects of Rainfall on Atmospheric Transmissivity

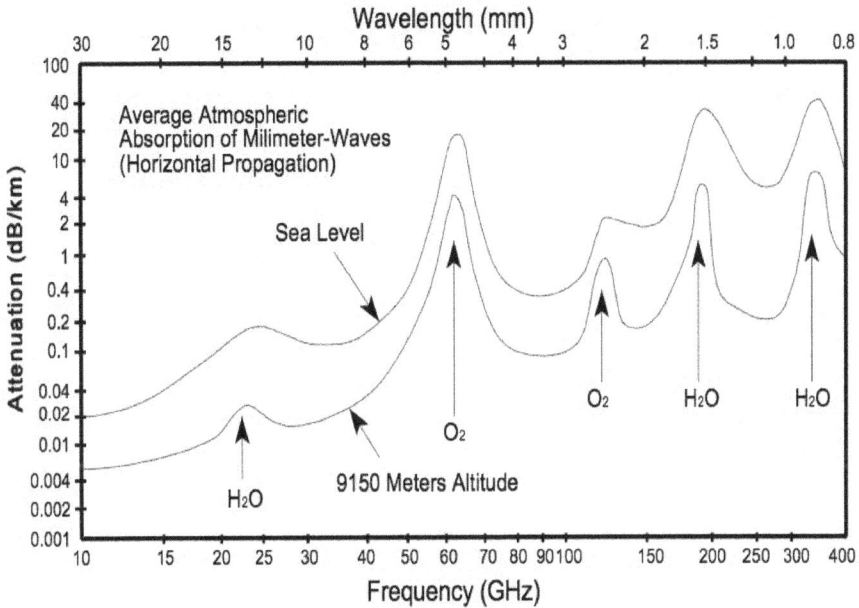

Atmospheric Transmissivity at Radio Frequencies showing attenuating properties of Oxygen and Water Vapor

Interstellar Path Loss

Ignoring the effects of Ionospheric, Water and Oxygen attenuation, we can look at the communications losses between stars up to 100 Light years away. We can do this at various frequencies and with different sized antennas. The losses are significant which makes this form of communications very challenging, even with the highest amount of transmitting power. Another requirement is a very large antenna, with high gain, to attempt to overcome the path losses. As it turns out the higher the gain of the antenna, the smaller the beam width. This is a problem as it also means that the field of view becomes smaller and thus it takes more time to examine a reasonable portion of the sky.

Path Loss in dB

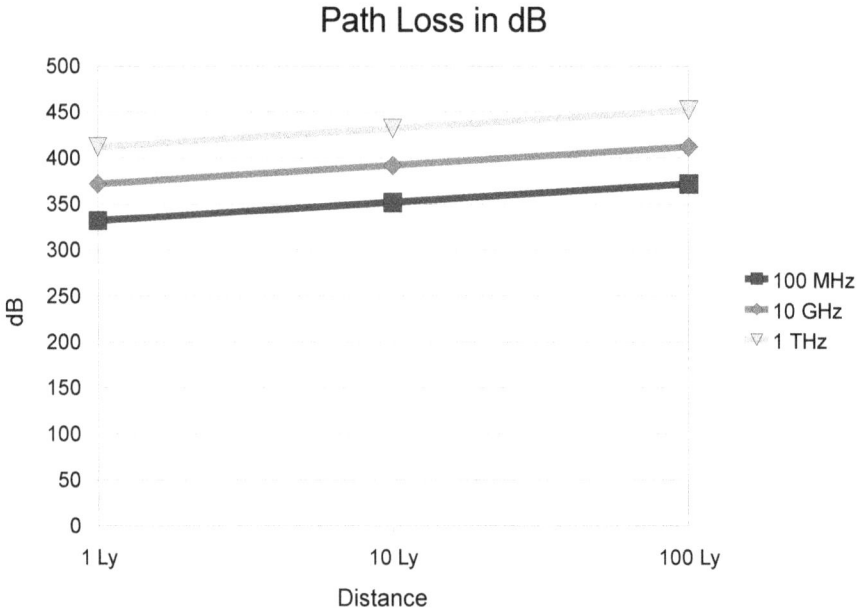

Path Loss over Interstellar Distances

As can be seen in the above graph, the losses are significant. Assuming that a receiver is capable of making up 100 dB of loss, this leaves at least 250 Db that has to be made up by antenna gain (both sides) and transmitter power levels. Even with huge antennas, the power level requirements of the transmitter far exceed anything that has ever been built.

Again, assuming very large antennas, the beam width would

necessarily be very small and thus the two communicating parties would have to be looking directly at the other planet at the time when the first planet's radio signals are predicted to appear, compensating for the speed of light (which radio waves travel at). Considering these long periods between transmission and reception, planets and solar systems move and thus predictions need to be made as to where to point the antennas as well.

Directivity vs. Frequency

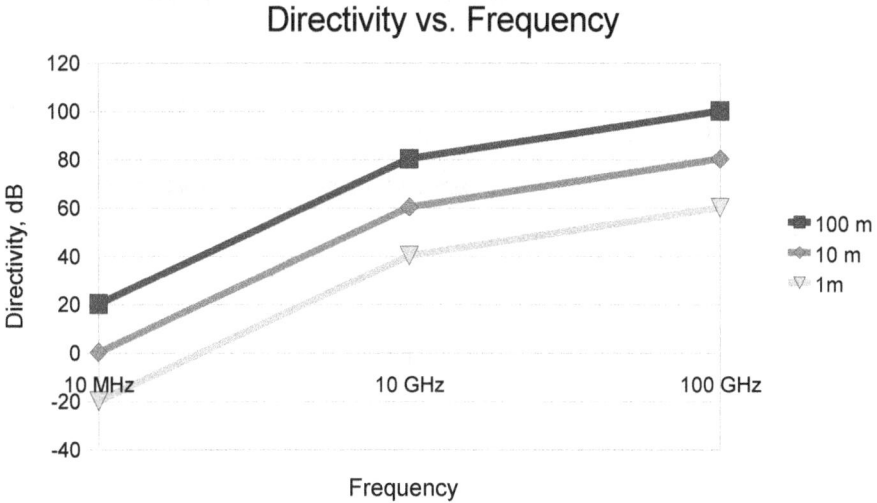

Antenna Gain (or Directivity) vs. Frequency for 3 Diameters

Large Antenna Issues

Another requirement is a very large antenna, with high gain, to attempt to overcome the path losses. As it turns out the higher the gain of the antenna, the smaller the beam width as mentioned before. The formula to calculate beam width is:

$$Beam\,width,\,Degrees = 57.3*\left(\frac{\lambda}{D}\right)$$

where

λ = wavelength or 300/MHz in meters

D = Diameter in same units as lambda

With a given antenna, the beam width also changes with frequency as frequency and wavelength are related by

$$lambda = \frac{c}{f}$$

where:

lambda = wavelength in meters

c = speed of light (300,000,000 meters per second)

f = frequency in Hertz

This means that the same antenna has a wide beam at lower frequencies and a narrower beam at higher frequencies. This also means that the same antenna has lower gain at the lower frequencies and higher gain at the greater frequencies.

During scans of the NASA SETI (or HRMS) program this was evident with the large radio telescopes they employed. The gain and beam width scales as the frequency, as seen below:

Gain vs. Beamwidth

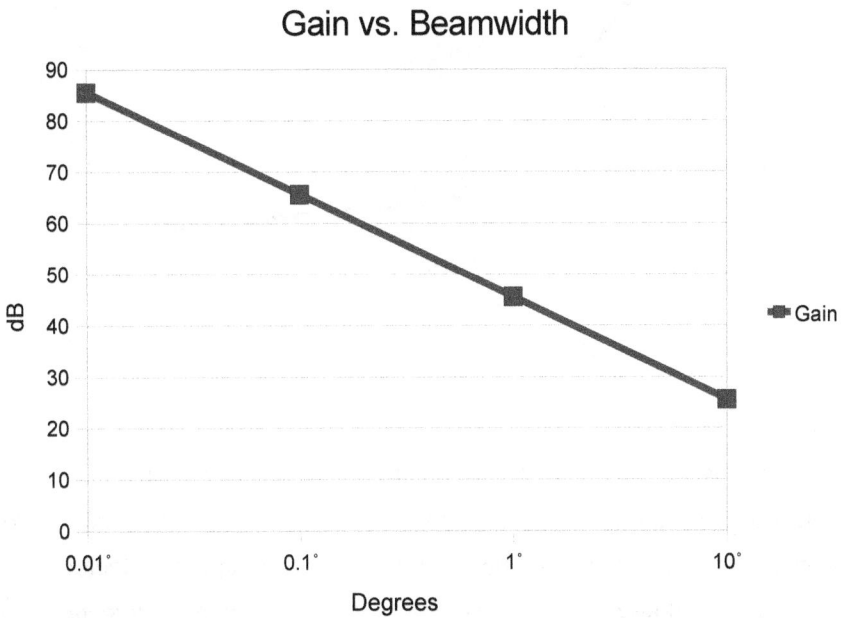

Beamwidth
Size vs. Frequency

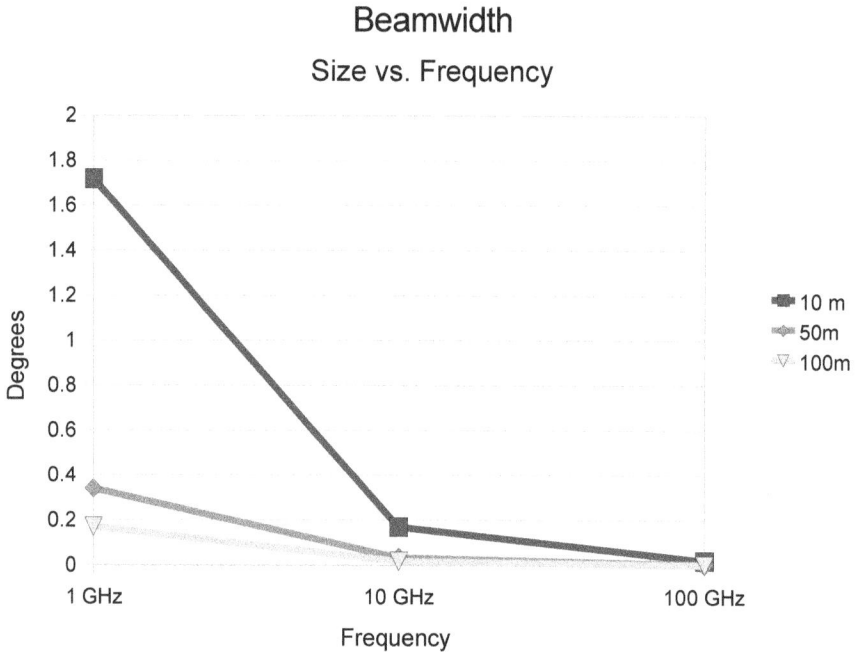

Reviewing the Radio Astronomy equation presented previously reveals that there are several ways to enhance detectability. One of the most popular approaches in astronomy is to integrate the signals detected. This means that only slowly changing features of the signals can be detected. This does however increase the sensitivity by the following factor:

$$Increase\ in\ Sensitivity = \sqrt{(N)}$$

where:

N = Number of Integration Periods

	Seconds	Increase in Sensitivity:
	1000	31.62
	100	10
Integration	10	3.16
Time:	1	1
	0.1	0.32
	0.01	0.1

Examples of Sensitivity Increases vs. Seconds

Transmitter Characteristics

Transmitters rarely operate at peak power, typically they are changing output power levels in response to modulation, where information is carried. There are some notable exceptions, FM radio for instance operates at full power but changes slightly in frequency to transfer information. TV and Radar broadcasts are usually at less than full power. The advantage of using less power is lowering the cost of the transmitter by using peak power levels infrequently. Radar for instance usually send out pulses of very high magnitude followed by quiet periods to let the transmitter electronics recover. The duty cycle of these types of pulse radar systems is usually less that 1%. This means that for less than 1% of the total time, the transmitter is on. Today's radar systems can however produce a significant amount of power, hundreds of thousands, sometimes millions of watts for these brief periods. This power is certainly enough to "light" up an object of interest. After these large pulses, radars listen for the echos and plot their amplitudes and ranges.

The following table shows the relationship of Peak vs. Average power levels. This has a consequence when trying to observe the radars or other transmitters from a very long distance away. Keep in mind that radars usually move their antennas to scan for multiple targets, thus allowing the receivers at a distance only a few hundred milliseconds to observe the pulses.

	Duty Cycle					
	100.00%	50.00%	25.00%	10.00%	1.00%	0.10%
	20 MW	10 MW	5 MW	2 MW	200 KW	20 KW
	10 MW	5 MW	2.5 MW	1 MW	100 KW	10 KW
	1 MW	500 KW	250 KW	100 KW	10 KW	1 KW
	500 KW	250 KW	125 KW	50 KW	5 KW	500 W
	100 KW	50 KW	25 KW	10 KW	1 KW	100 W
Average	50 KW	25 KW	12.5 KW	5 KW	500 W	50 W
Power	10 KW	5 KW	2.5 KW	1 KW	100 W	10 W
	5 KW	2.5 KW	1.25 KW	500 W	50 W	5 W
	1 KW	500 W	250 W	100 W	10 W	1 W
	500 W	250 W	125 W	50 W	5 W	.5 W
	100 W	50 W	25 W	10 W	1 W	.1 W
	10 W	5 W	2.5 W	1 W	.1 W	01 W
	5 W	2.5 W	1.25 W	.5 W	.05 W	.005 W
	1 W	.5 W	.25 W	.1 W	.01 W	.001 W

Peak vs. Average Power

Life in the Universe and Where to find it

What We have Transmitted to the Stars

One of the more important issues that come out of scientists and science fiction writers is an understanding of what we have transmitted to the cosmos over the history of broadcast. These transmissions include spark gap to laser and have increasingly higher outputs over the years. During the first years of radio broadcasts the power levels were very low and "dirty" meaning that the purity of the signals was very low. The Titanic's famous SOS for instance was broadcast around 1 MHz, now the center of our AM band, and had an output level of around 200 watts. Today we track missiles with radars with output power exceeding 20 Million Watts at significantly higher frequencies.

The Earth has transmitted energy for radar and communication usually starting at low frequencies and moving higher and higher as technology permits. As we did so, the antennas became more sophisticated and more directional in operation.

Keep in mind that transmissions below 30 MHz can be heavily attenuated by the Ionosphere and that transmissions above 10 GHz can also be attenuated by water vapor and oxygen in the atmosphere. In the so called "water window" of between 1 and 10 GHz, the predominant form of broadcast is radar, satellite and point to point microwave. In these cases, directional high gain antennas are used to improve link margins. This of course minimizes any radiation to unintended areas, like the stars. In the case of radars, whose beams typically rotate at a predicable rate in the horizontal plane, there is a chance of observing their signals if they are tangential to the receiver out in space. Atmospheric ducting in these cases can still attenuate their levels.

172

For TV transmissions, the transmitting antennas usually have shaped beams to best cover their audiences. There is no point in transmitting to the stars. Cellular towers employ the same techniques. As a result the idea of observing television programs on Vega that are sent from Earth would be near impossible, both because we do not intentionally aim our antennas to Vega and because of the relatively low power levels.

We have transmitted radio, TV, radar, microwave and optical signals to space. This is rarely on purpose. Signals to control satellites are usually of medium power with large antennas and thus narrow fields of view to avoid overlapping other satellites. The return signals are moderate in power to minimize weight and energy usage on the spacecraft. They also point their antennas directly at Earth to cover specific areas.

By far the largest transmission we have sent to the stars has been the Electro Magnetic Pulse (EMP) from nuclear explosions. Some high altitude versions of Mega ton bombs had enough energy to blow out light bulbs thousands of miles away. These high altitude explosions radiated omni-spherically and certainly had the best chance of detection on another star system. There was of course, no modulation, just a very intense pulse estimated at 30,000 volts/meter at 100 miles for 1 uSec. Perhaps this is a method to detect other stars' nuclear activities.

	Spark Gap	Vacuum Tube	Magnetron Klystron TWT	Analog Modes	Digital Modes	Wideband Modes
Year	1887-1916	1930	1940	1960	1980	2010
Frequency of Operation	.545 – 1.2 MHz	.500 – 30 MHz	.500 – 10 GHz	.500 – 20 GHz	.500 – 60 GHz	.500 – 1 THz
Emitter Type	Titanic	Broadcast	Radar	Communications	Satellite	Internet
Transmission Source			Hiroshima Nagasaki	Starfish Prime Soviet Test 184	Troposcatter DEW Line	High Energy Laser
Maximum Propagation Distance	110 Ly	80 Ly	70 Ly	50 Ly	30 Ly	1 Ly
# Stars within Reception distance	>14508 Stars	7428 Stars	4976 Stars	1813 Stars	391 Stars	0 Stars
Peak Power in Watts	200 Watts	1000 Watts	100,000 Watts	10,000,000 Watts	1,000,000 Watts	2,000,000 Watts
Peak Power in dB	53 dB	60 dB	80 dB	100 dB	90 dB	93 dB

Chronology of Earth's Transmissions

Types of Transmitters in History

Spark gap (wide band, noisy) designed by Tesla and Marconi, is usually made up of a voltage source modulated with a telegraph key, driving a high voltage resonant circuit. The spark gap (7) is driven by a high voltage coil, in this case from a Ford Model T automobile. The output of the coil and resonant circuit drive the antenna are shown in the background. Transmissions were low frequency and wide band. Operations using this type began around 1910.

Above is a 20 KHz transmitter used for Very Low Frequency (VLF) transmissions. This is actually an alternator which physically rotates at 20,000 RPM. The output sinusoid wave form drives an antenna which in this case covered many miles. It is modulated with a telegraph key and is still in use today, where it comes on air once a year typically in the Fall. The low frequency allows for very good propagation around the Earth (not very well to space) and was the progenitor for low frequency submarine communications and navigation systems like LORAN. The quality of world wide propagation allowed the design of several oceanic navigation systems during and after the second world war, LOng RAnge Navigation (LORAN) was one such system and only recently went off air.

Another low frequency of interest is the 50 and 60 Hz of AC power that courses across the entire world to provide energy to billions of people. This emission of course is not modulated.

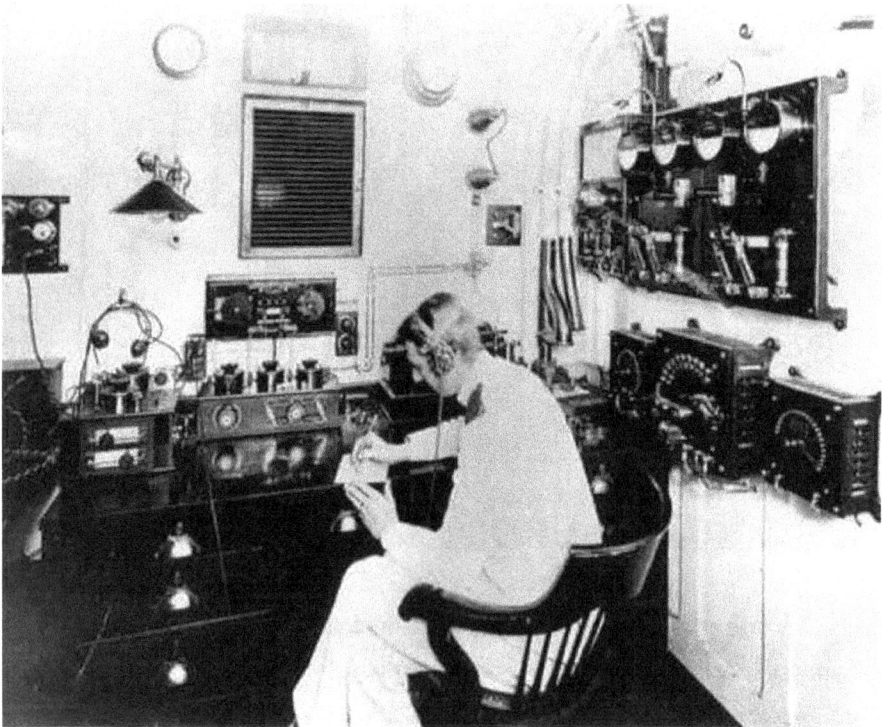

The Radio Room on the Titanic

On the bridge of the Titanic a radio operator used a spark gap transmitter to send messages, typically telegrams to people all over the world. It was also used to send the SOS message when the ship was struck an iceberg in the Northern Atlantic on April 14, 1912. The signal was heard by other ships and shore stations many thousands of miles away. The distance between Titanic and the closest ship in the Atlantic prevented a speedy rescue operation and many people were lost at sea as a result. This transmitter put out several hundred watts of power and was connected to a large antenna system above decks. It operated at approximately .5 and 1 MHz.

The second world war (WWII) saw the quick development of shortwave communication as well as the advent of radar. Shortwave took advantage of the reflective properties of the ionosphere and operators learned to operate below the maximum usable frequency (MUF) during the course of the day. This was not conducive to radiating energy to the stars. Typically lower frequencies were used during the evening and higher one during the day. Shortwave communication are still used extensively today and accurate propagation prediction maps are updated on a 5 minute basis. The communications are still used by ships, radio amateurs and the military.

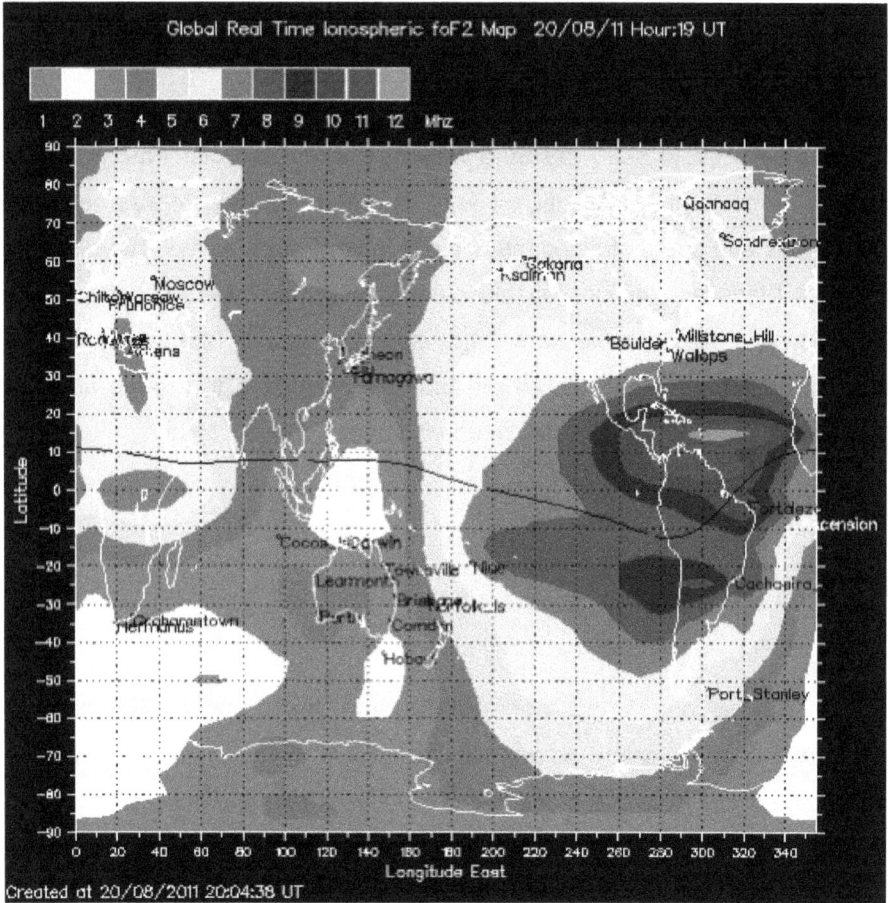

Recent MUF chart

Also, during WWII, radars were developed to follow the movements of planes and aircraft. Several opposing countries had this technology and used it at various frequencies to track their enemy. Low frequency in the VHF bands were used initially, prohibiting the use of this powerful tool in aircraft.

With the invention of the cavity magnetron, airborne applications were developed and could be used for tracking other aircraft as well as locating submarines.

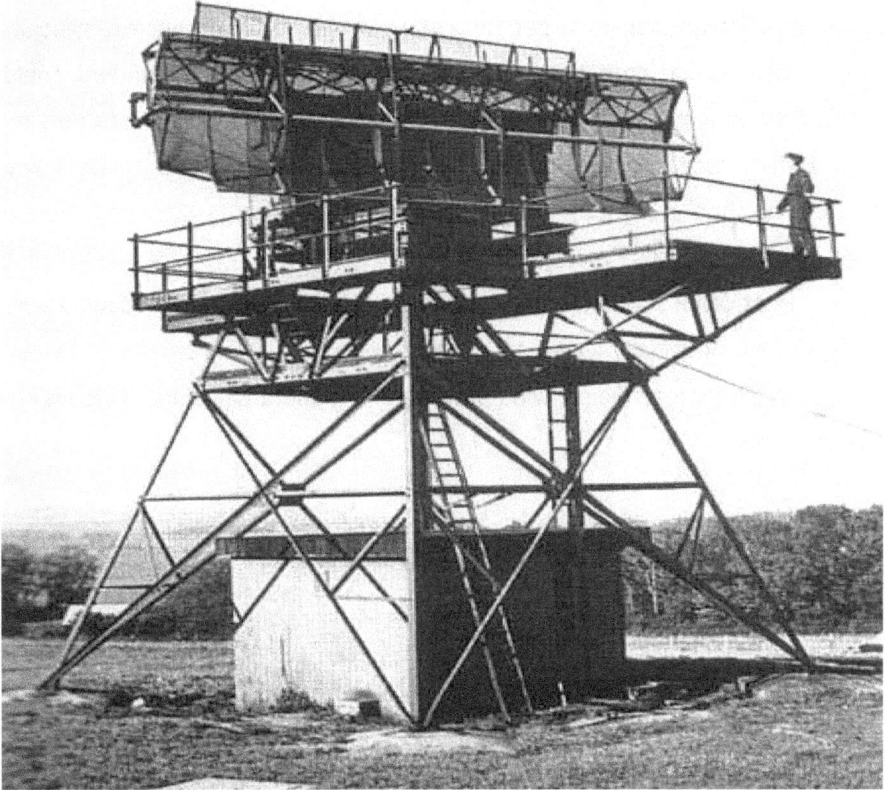

British Radar Installation in WWII

One of the First Radars used in England during WWII was this system which used an azimuth scanning antenna that created a "fan" beam to cover large vertical angles as it rotated. These radar systems were typically placed near the shore lines to watch for enemy planes and were instrumental in winning the Battle of Britain. The operators who viewed the radar scopes sent their target information to a central station where several people physically plotted them on a large horizontal map of the area. Commanders could observe the enemy's flights and send aircraft for interception. Another version of a British radar system is the Chain Home which used much lower frequencies than the radar shown above and used large rotating antenna towers.

Post WWII Radar technical research was intense and produced a very wide variety of radar systems. These were used for aircraft, weather, navigation, sea state studies, wind profiling and planetary (including the Moon) observations.

Altair Radar on the Kwajalein Atoll used for tracking missiles and other space items. It has a power output of 10 Million Watts in the VHF and 20 Million Watts in the UHF range

Radar Astronomy

Beginning in the 1960s, technology had advanced enough to be able to send a signal to a planet and receive its reflection. The surfaces of Venus, Mars and the Moon were examined in detail, leading to some of the first high resolution radar images of their respective surfaces. Digital modulation techniques were also developed to achieve both higher average power and high range resolution. The antennas at Arecibo, the Deep Space Network and the larger radio telescopes in the world were employed for this activity. As Venus had an optically impenetrable cloud layer, this techniques was particularly useful.

70 Meter Radar Telescope in Russia

Using extremely stable frequency sources, several observatories have been used simultaneously to affect a method to produce very high resolution of many of the planets, asteroids and moons in the solar systems. This is the technique of large baseline interferometry, where a particular radar telescope transmits a high power signal towards a stellar object and one or more radio telescopes or radio telescope arrays receive the signals. Due to their separation, the resolution of an antenna equivalent to the maximum telescope to telescope distance can be observed.

Using commercially available equipment, radio amateurs have been frequently reflecting signals off of the surface of the moon to make contact with other amateurs at great distances. This is known as Earth Moon Earth (EME) communications. There is a measurements delay due to the Moon's distance. This distance varies with the motions of the moon and can be measured with great accuracy. These communication links are typically done at VHF and UHF bands where the ionosphere has little effect.

The Very Large Array in Socorro, New Mexico
used as a receive site for radar astronomy

NASA's Deep Space Network (DSN) consists of three 73 meter telescopes that have been referred to many times as the most sensitive radio telescopes in the world. These facilities are used for communicating to deep space probes as well as performing radar telescope duties. The dishes are located in California, Spain and Australia and thus can keep a constant link with space craft that have traveled through and beyond our solar system.

73 Meter Deep Space Network Station in Goldstone, California
used for interplanetary missions as well as radar astronomy

Life in the Universe and Where to find it

Radio Broadcast

Today thousands of transmitters send music, television and other programming to millions of people on Earth. The vast majority are in the AM band (550 KHz – 1.7 MHz) and the FM band (88 MHz to 108 MHz) . The AM band stations lower their power outputs at sundown to minimize "over" propagation where their signals could interfere with distant stations operating on the same frequency. During the evening the ionosphere become more reflective in the AM band allowing for some stations to heard for over a thousand miles that would normally only be only heard over a few hundred.

High power AM transmitter 2.5 Million Watts

The antenna systems for AM, FM and TV broadcasting are typically designed to emit in an omni-directional patterns with low elevation angles. This is meant to cover the broadcast area in the most efficient manner. Some of these antennas are in array form to make a more directional pattern, again for most efficient coverage.

189

TV Broadcast

Beginning in the 1930s, many countries began broadcasting video images to major cities. The video is in the form of amplitude modulation and the accompanying voice channel in modulated with frequency modulation. The signals are usually high power, on the order of 100s of thousands of watts with directional antennas. The TV channels start at around 55 MHz and go up to about 700 MHz. All of these frequencies are minimally effected by the ionosphere and can propagate into space although the downtilt of the transmitting antenna patterns precludes them from doing so.

A High Power TV transmitter

TV broadcast antennas tend to have patterns customized for their viewing audience and like FM broadcast, can have circular polarity to enhance signal propagation. An attractive feature of these broadcasts is a periodic sync frequency that by itself makes detection easier although again, not at interstellar distances.

Military Transmissions

After the second World War and the advent of the Cold War, there were significant advancements in radar, satellite communications and point to point microwave communications. In addition, directed energy weapons were developed using high power microwave, laser and even X-ray radiators. High power surveillance radars were designed and placed along a line in Northern Canada called the Distant Early Warning (DEW) line then high power phased arrays were placed on the East and West coast of the United States for the purposes of over the horizon detection of aircraft and ships. Russia had similar systems. These radars had power outputs exceeding 1 Million watts. Many of the radars had troposcatter communications links to transfer the detected aircraft positions to a central base. These troposcatter systems used tens of thousands of Watts and relied on the reflection of microwave signals off of the troposphere. There are dozens of radars and troposcatter systems across the Canadian Arctic, from Alaska to Maine developed and installed from approximately 1950 to present.

Troposcatter and high power radar on DEW line

Details of Troposcatter antenna and feed

These radar and communications systems were augmented by satellite uplinks and Airborne Warning And Control (AWAC) type surveillance systems, creating a very noisy environment as far as radio emissions are concerned. Although most of the energy is directed horizontally, much of it has escaped and could be monitored from a distant place. The challenge would be in the fact that most of the radars scanned in azimuth give very little energy in any one particular spot. The point to point tropospheric links are fixed but relied on the reflection of atmospheric phenomena and thus would

193

not transmit much information into space.

Troposcatter systems work in the microwave frequencies between 2 and 5 GHz, they used enough power to guarantee reliable communications and almost always use some form of encryption.

Over the Horizon Backscatter Radar and HAARP

During the Cold War, an interesting radar technology was developed to overcome the line of sight problems of standard radar systems. In the microwave region, radars have a useful range of 0 to 250 miles and at the longer ranges, low altitude targets cannot be seen due to the curvature of the Earth. This problem necessitates multiple radar systems with minimal separation to get the coverage required. The idea came up to use the multiple skip phenomena of shortwave communications to locate very distant targets, initially ships at sea. The challenge was to be able to separate the noise of propagation and sea clutter from the ships. The challenges were overcome and several systems were designed and built, with successful results. These are known as Over the Horizon Backscatter (OTH-B) radar systems. There are still several of the systems in operation today typically based near a sea coast. They operate with power levels up to 10 Million Watts and in the case of shortwave systems, between 5 and 28 MHz. There are a few units that run at much higher frequencies but without the same range of operations. The shortwave variety can see several thousand miles out to sea. Shortwave listeners and Amateur Radio operators sometime here these transmissions, especially is they are swept in frequency. They sound like birds chirping when listening in the band of operation.

194

Life in the Universe and Where to find it

The High Frequency Active Auroral Research Program or HAARP is an ionospheric radar system in Alaska that operates between 7.4 and 9.4 MHz with very high power, enough to heat the ionosphere. The antenna array looks vertically and is intended to probe and understand the workings of the ionosphere.

Over the Horizon Backscatter Radar Antenna (OTH-B)

EMP from Nuclear Weapons

Electromagnetic Pulse (EMP) is defined as a burst of electromagnetic radiation which is typically generated by a nuclear explosion or a sudden fluctuation in a magnetic field. The resulting changing electric fields and magnetic fields can couple with electronic equipment to produce damaging voltage and current surges. These surges can disable electrical systems such as radios, computers, lights and can disable vehicles.

In July of 1962 the United States detonated a nuclear warhead 250 miles above the surface of the Earth, in the South Pacific. This test was known as Starfish Prime and had a significant amount of EMP output, enough to cause electrical damage in Hawaii, some 900 miles distant. Streetlights were blown out, burglar alarms set off and microwave links damaged.

This interaction of the very rapidly moving negatively charged electrons with Earth's magnetic field radiates a pulse of electromagnetic energy. The pulse typically rises to its peak value in about 5 nanoseconds. The magnitude of this pulse typically decays to half of its peak value within 200 nanoseconds. During nuclear tests in 1962, EMP disruptions were suffered aboard KC-135 photographic aircraft flying 190 miles from the 410 Kiloton (1,700 TeraJoule) Bluegill Triple Prime and similar yield Kingfish detonations at 30 and 59 mile burst altitudes, respectively but the vital aircraft electronics were far less sophisticated than today and the aircraft were able to land safely. [1]

The short duration of the EMP pulses creates a very brief signal at the frequency commensurate with the pulse length. For instance a 5 nSec pulse length corresponds to a frequency of 200 MHz. The pulse is of such intensity that those listening from many light years away should be able to hear it. This assumes they have very sensitive receivers and very large antennas that happen to be pointed at the Earth at the time of the explosion. From a philosophic point of view, this could be how aliens discover that the Earth has produced some very large weapons. There is of course a very intense short duration flash that is associated with the explosion as well. Optical astronomers from light years away should be able to see this flash. Just as Maxwell (and Heaviside) predicted, the fluctuating of a magnetic field can produce radio emissions which EMP and other cosmic forces are capable of doing. Gamma rays are also produced in this process as depicted in the following graph.

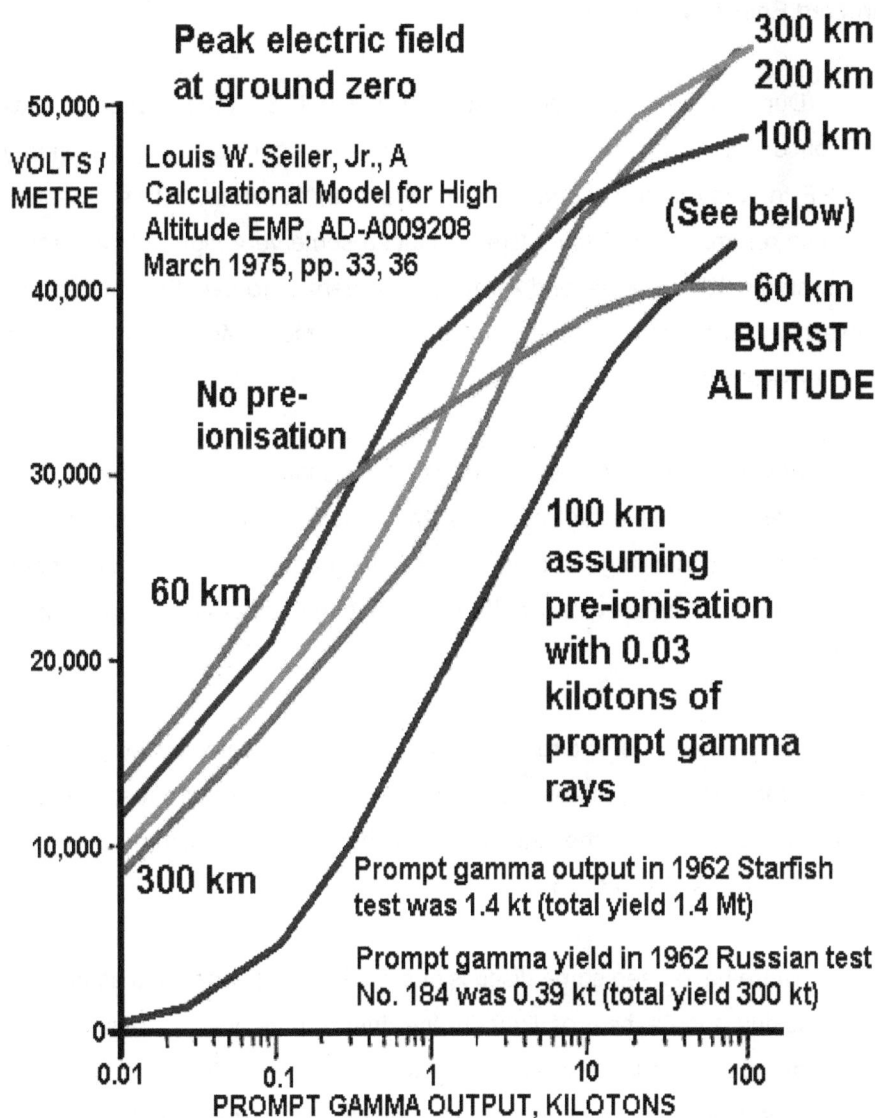

Peak electric field at ground zero

Louis W. Seiler, Jr., A Calculational Model for High Altitude EMP, AD-A009208 March 1975, pp. 33, 36

No pre-ionisation

60 km

300 km

300 km
200 km
100 km

(See below)

60 km
BURST ALTITUDE

100 km assuming pre-ionisation with 0.03 kilotons of prompt gamma rays

Prompt gamma output in 1962 Starfish test was 1.4 kt (total yield 1.4 Mt)

Prompt gamma yield in 1962 Russian test No. 184 was 0.39 kt (total yield 300 kt)

VOLTS / METRE

PROMPT GAMMA OUTPUT, KILOTONS

EMP Effects

Directed Energy Weapons

During the Reagan administration in the U.S., much research was done on directed energy weapons, or those that could concentrate energy from RF to X-ray in a focused pattern toward an adversary. This research, sometime referred to as "Star Wars" produced some very high output infrared laser systems, RF systems, and X-ray laser systems. To date the Infrared and Visible high output lasers have been developed into weapons for defensive and offensive purposes.

Power levels into the hundreds of thousands of watts have been reported, although the actual numbers are classified. These weapons have been placed onboard large aircraft and are being adapted for smaller aircraft. No doubt they will be prevalent onboard Unmanned Aerial Vehicles (UAVs) in the near future.

These weapons, if directed skyward, could be detected by people in close in star systems. This assumes that they have powerful telescopes and happen to be looking in the right place at the right time, compensating for propagation and stellar motion.

There have been some discussions within the astronomical community about looking for flashes of light in the Infrared as well as visible light spectrums. Covering large areas of sky with the required telescope power is the challenge however. As mentioned before, the larger the telescope or antenna, the smaller the field of view and thus to have a single large telescope cover the entire sky would very time consuming.

Boeing 747 with high power chemical laser aboard. Beam exits nose turret

An experimental X-Ray laser

Doubtless, it would be ambitious to try to monitor the whole sky at all wavelengths for signs of flashes indicative of directed energy weapons. However, several fortuitous observations in the radio spectrum have detected pulses, some long in duration, emanating from the cosmos. Kraus found one in the 1970s and the Deep Space Exploration Society found one in the 1990s. The question is whether or not these were generated by normal physical processes or from intelligent life.

Arrays like the "Shoemarkian" system are capable of observing these large areas of sky but need to be developed for the particular portions of the electromagnetic spectrum.

High Intensity Millimeter Radar system, 70 KiloWatt output peak, 94 GHz with 62.5 dBi Antenna gain

This radar and the ones mentioned above put out large amplitude pulses of a defined length and period. These factors create a detectable pattern that can be observed from a great distance if one has the knowledge and equipment to look for it. So to are the characteristics of the high energy lasers.

The problem of course is that we presuppose that this knowledge and equipment is on the natural course of evolution for other interstellar life. That might not be the case, actually, it *will* not be the case. We will discuss this in detail later.

Emissions from Satellites

Historically, the evolution of satellite technology required that they put out lower power levels first before reliable high output transmitters were available and capable of operations in the challenging environment of space over many years. The standard requirement is that a satellite used for commercial broadcasting is that is works for 17 years without replacement. These satellites are typically placed in geosynchronous orbit, some 23,500 miles above the Earth's surface so they appear stationary in the sky. This is because they are moving through space at exactly the same rate as the Earth is rotating. These satellites concentrate their power over specific fields of view, all on the surface of the planet. Doing so minimizes or eliminates any radio emissions to the sky. As such, these are not good candidates for interstellar beacons. In addition, the power levels are usually modest, this is based on available transmitting equipment that can stand the rigors of space for a long period of time.

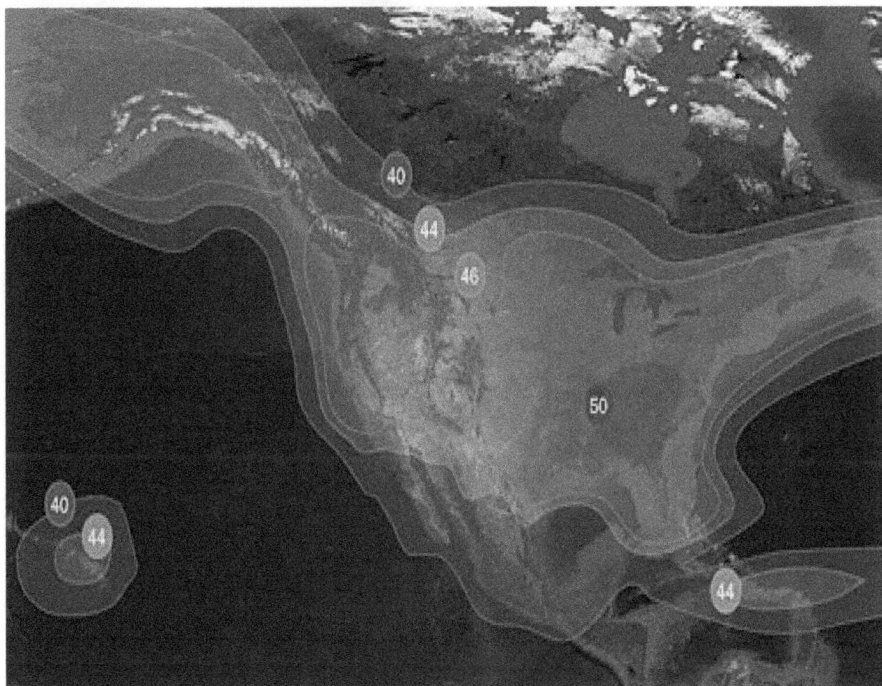

A typical communication satellite's footprint showing dB contours on the surface

Other satellites that may put out more power, for instance military designs, are typically encrypted and require knowledge of the encoding scheme to be detectable.

Conclusion

The distances between the Earth and even the closest stars precludes any chance of receiving sentient produced radio transmissions.

The transmissions from Earth over history are too weak to be heard beyond about one light year away from our own solar system.

Chapter 9 - The Discovery of Extra-Solar Planets

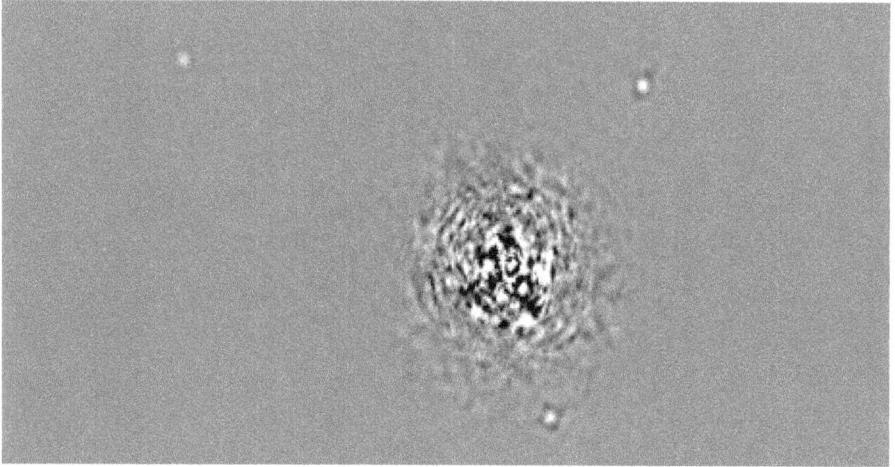

Three planets around star HR 8799 taken by Keck with adaptive optics

The first published, confirmed discovery was made in 1988 by the Canadian astronomers Bruce Campbell, G. A. H. Walker, and S. Yang. Although they were cautious about claiming a planetary detection, their radial-velocity observations suggested that a planet orbited the star Gamma Cephei. Partly because the observations were at the very limits of instrumental capabilities at the time, widespread skepticism persisted in the astronomical community for several years about this and other similar observations. Another source of confusion was that some of the possible planets might instead have been brown dwarfs, objects that are intermediate in mass between planets and stars. The following year, however, additional observations were published that supported the reality of the planet orbiting Gamma Cephei, though subsequent work in 1992 raised serious doubts.Finally, in 2002, improved techniques allowed the planet's existence to be confirmed. [1]

In early 1992, radio astronomers Aleksander Wolszczan and Dale Frail announced the discovery of planets around another pulsar, PSR 1257 + 12. This discovery was confirmed, and is generally considered to be the first definitive detection of exoplanets. These pulsar planets are believed to have formed from the unusual remnants of the supernova that produced the pulsar, in a second round of planet formation, or else to be the remaining rocky cores of gas giants that survived the supernova and then decayed into their current orbits.

On October 6, 1995, Michel Mayor and Didier Queloz of the University of Geneva announced the first definitive detection of an exoplanet orbiting an ordinary main-sequence star 51 Pegasi. This discovery, made at the Overvatoire de Haute-Provence, ushered in the modern era of exoplanetary discovery. Technological advances, most notably in high-resolution spectroscopy, led to the detection of many new exoplanets at a rapid rate. These advances allowed astronomers to detect exoplanets indirectly by measuring their gravitational influence on the motion of their parent stars. Additional extrasolar planets were eventually detected by observing the variation in a star's apparent luminosity as an orbiting planet passed in front of it. [1]

As of August 10, 2011, 573 confirmed exoplanets are listed in the *Extrasolar Planets Encyclopaedia,* including a few that were confirmations of controversial claims from the late 1980s. The first system to have more than one planet detected was PSR 1257 + 12; the first confirmed to have multiple planets orbiting a main-sequence star was Upsilon Andromedae. 60 such multiple-planet systems are known as of April 2011. Among the known exoplanets are four pulsar planets orbiting two separate pulsars. Infrared observations of circumstellar dust disks suggest millions of comets in several extrasolar systems. [1]

Planets are extremely faint light sources compared to their parent stars. At visible wavelengths, they usually have less than a millionth of their parent star's brightness. It is difficult to detect such a faint light source, and furthermore the parent star causes a glare that tends to wash it out.

A team of researchers from NASA's Jet Propulsion Laboratory demonstrated a technique for blocking a star's light with a vector vortex coronagraph, thus enabling direct detections to be made more easily. The researchers are hopeful that many new planets may be imaged using this technique. Another promising approach is nulling interferometer.

At the moment, however, the vast majority of known extrasolar planets have only been detected through indirect methods. The following are the indirect methods that have proven useful:

Radial velocity or Doppler method

As a planet orbits a star, the star also moves in its own small orbit around the system's center of mass. Variations in the star's radial velocity— that is, the speed with which it moves towards or away from Earth— can be detected from displacements in the star's spectral lines due to theDoppler effect. Extremely small radial-velocity variations can be observed, down to roughly 1 m/s. This has been by far the most productive method of discovering exoplanets. It has the advantage of being applicable to stars with a wide range of characteristics.

Transit method

If a planet crosses (or transits) in front of its parent star's disk, then the observed brightness of the star drops by a small amount. The amount by which the star dims depends on its size and on the size of the planet, among other factors. This has been the second most productive method of detection, though it suffers from a substantial rate of false positives and confirmation from another method is usually considered necessary.

Transit Timing Variation (TTV)

TTV is a variation on the transit method where the variations in transit of one planet can be used to detect another. The first planetary candidate found this way was exoplanet WASP-3c, using WASP-3b in the WASP-3 system by Rozhen Observatory, Jena Observatory, and Torun Centre for Astronomy The new method can potentially detect Earth sized planets or exomoons.

Gravitational microlensing

Microlensing occurs when the gravitational field of a star acts like a lens, magnifying the light of a distant background star. Planets orbiting the lensing star can cause detectable anomalies in the magnification as it varies over time. This method has resulted in only a few planetary detections, but it has the advantage of being especially sensitive to planets at large separations from their parent stars.

Astrometry

Astrometry consists of precisely measuring a star's position in the sky and observing the changes in that position over time. The motion of a star due to the gravitational influence of a planet may be observable. Because that motion is so small, however, this method has not yet been very productive at detecting exoplanets.

Pulsar timing

A pulsar (the small, ultra dense remnant of a star that has exploded as a supernova) emits radio waves extremely regularly as it rotates. If planets orbit the pulsar, they will cause slight anomalies in the timing of its observed radio pulses. Four planets have been detected in this way, around two different pulsars. The first confirmed discovery of an extrasolar planet was made using this method. Most pulsars radiate maximum radiation in the 200 – 600 MHz band.

Timing of eclipsing binaries

If a planet has a large orbit that carries it around both members of an eclipsing double star system, then the planet can be detected through small variations in the timing of the stars' eclipses of each other. As of December 2009, two planets have been found by this method.

Circumstellar disks

Disks of space dust surround many stars, and this dust can be detected because it absorbs ordinary starlight and re-emits it as infrared radiation. Features in the disks may suggest the presence of planets.

Most extrasolar planet candidates were found using ground-based telescopes. However, many of the methods can work more effectively with space-based telescopes that avoid atmospheric haze and turbulence. COROT (launched December 2006) and Kepler (launched March 2009) are the two currently active space missions dedicated to searching for extrasolar planets. Hubble Space Telescope and MOST have also found or confirmed a few planets. The Gaia mission, to be launched in March 2013, will use astrometry to determine the true masses of 1000 nearby exoplanets. [1]

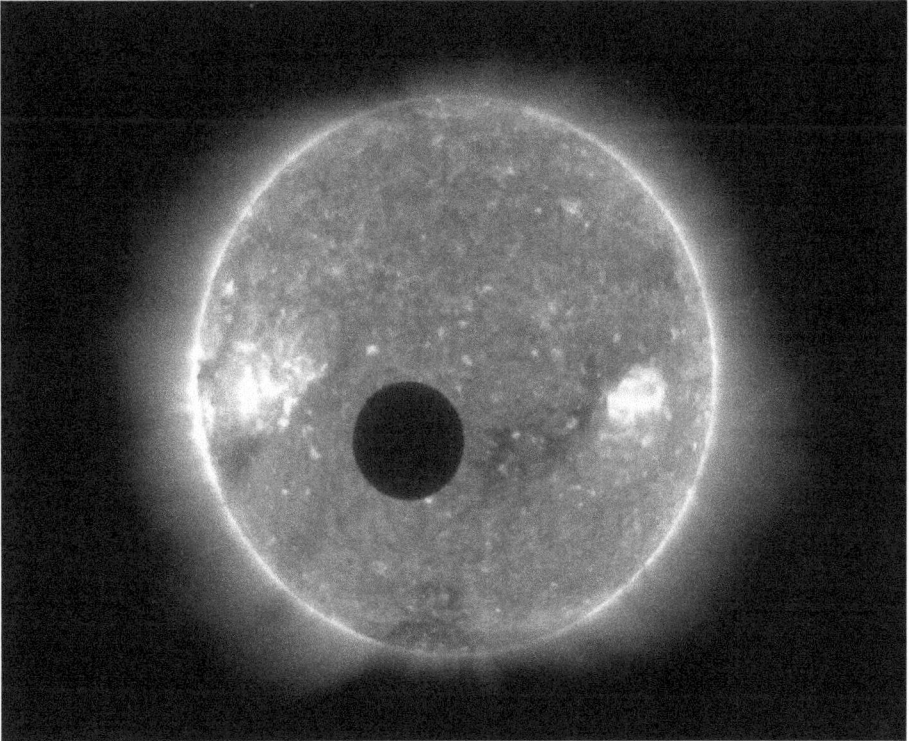

Transit of Moon over our Sun

Number of stars with Planets

Most of the discovered extrasolar planets lie within 300 light years of the Solar

System.

Planet-search programs have discovered planets orbiting a substantial fraction of the stars they have looked at. However, the total fraction of stars with planets is uncertain because of observational selection effects. The radial-velocity method and the transit method (which between them are responsible for the vast majority of detections) are most sensitive to large planets on small orbits. For that reason, many known exoplanets are "hot Jupiters": planets of roughly Jupiter-like mass on very small orbits with periods of only a few days. It is now known that 1% to 1.5% of sunlike stars possess such a planet, where "sunlike star" refers to any main-sequence star of spectral classes F, G, or K without a close stellar companion. It is further estimated that 3% to 4.5% of sunlike stars possess a giant planet with an orbital period of 100 days or less, where "giant planet" means a planet of at least thirty Earth masses.

The fraction of stars with smaller or more distant planets remains difficult to estimate. Extrapolation does suggest that small planets (of roughly Earth-like mass) are more common than giant planets. It also appears that planets on large orbits may be more common than ones on small orbits. Based on such extrapolation, it is estimated that perhaps 20% of sunlike stars have at least one giant planet while at least 40% may have planets of lower mass.

Regardless of the exact fraction of stars with planets, the total number of exoplanets must be very large. Since our own Milky Way Galaxy has at least 200 billion stars, it must also contain billions of planets if not hundreds of billions of them. [1]

Life in the Universe and Where to find it

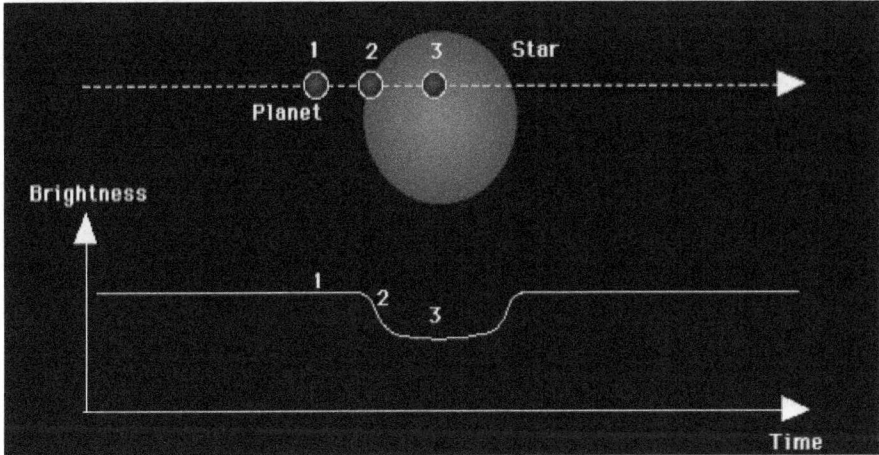

Transit of Extra-Solar Planet and resultant measurement

Characteristics of planet-hosting stars

The Morgan-Keenan spectral classification are numbers that follow letters OBAFGKM, which can be memorized by remembering "Oh be a fine girl, kiss me." The spectral class is represented on the Hertzsprung-Russell diagram where spectral class is plotted against luminosity. What becomes obvious is a wide swath of stars known as the "main sequence." What might not be so obvious is that stars move about in a predictable manor within this diagram during their life cycle.

Within this diagram all stars can be plotted, showing not only the "average" stars but also the giants, supergiants and white dwarfs. It has been immeasurably useful to astronomers in understanding the life cycles of stars.

212

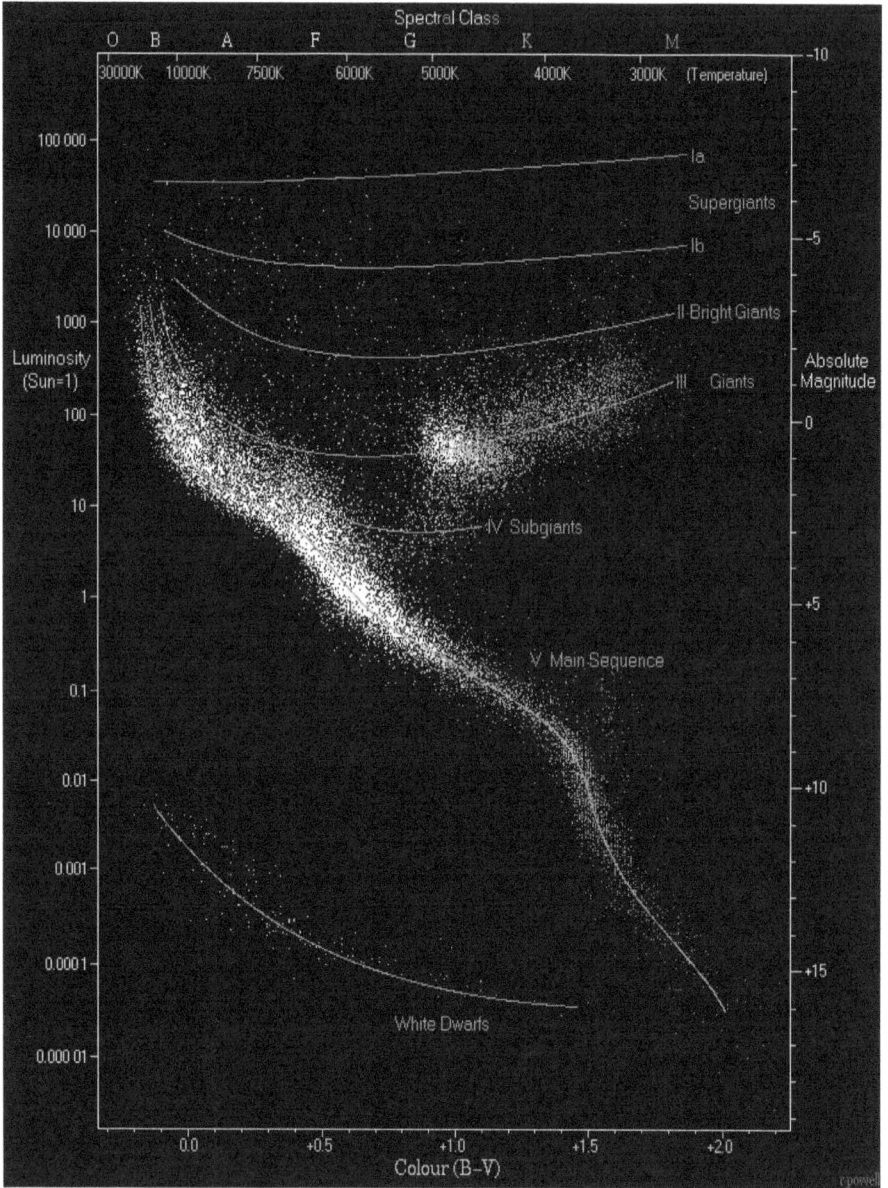

The Hertzsprung–Russell diagram

Class	Temperature (kelvins)	Conventional color	Apparent color	Mass (solar masses)	Radius (solar radii)	Luminosity (bolometric)	Hydrogen lines	Fraction of all main sequence stars
O	≥ 33,000 K	blue	blue	≥ 16 $M_\odot$	≥ 6.6 $R_\odot$	≥ 30,000 $L_\odot$	Weak	~0.00003%
B	10,000–33,000 K	blue to blue white	blue white	2.1–16 $M_\odot$	1.8–6.6 $R_\odot$	25–30,000 $L_\odot$	Medium	0.13%
A	7,500–10,000 K	white	white to blue white	1.4–2.1 $M_\odot$	1.4–1.8 $R_\odot$	5–25 $L_\odot$	Strong	0.6%
F	6,000–7,500 K	yellowish white	white	1.04–1.4 $M_\odot$	1.15–1.4 $R_\odot$	1.5–5 $L_\odot$	Medium	3%
G	5,200–6,000 K	yellow	yellowish white	0.8–1.04 $M_\odot$	0.96–1.15 $R_\odot$	0.6–1.5 $L_\odot$	Weak	7.6%
K	3,700–5,200 K	orange	yellow orange	0.45–0.8 $M_\odot$	0.7–0.96 $R_\odot$	0.08–0.6 $L_\odot$	Very weak	12.1%
M	≤ 3,700 K	red	orange red	≤ 0.45 $M_\odot$	≤ 0.7 $R_\odot$	≤ 0.08 $L_\odot$	Very weak	76.45%

Most known exoplanets orbit stars roughly similar to our own Sun, that is, main-sequence stars of spectral categories F, G, or K. One reason is simply that planet search programs have tended to concentrate on such stars. But even after taking this into account, statistical analysis indicates that lower-mass stars (red dwarfs, of spectral category M) are either less likely to have planets or have planets that are themselves of lower mass and hence harder to detect. Recent observations by the Spitzer Space Telescope indicate that stars of spectral category O, which are much hotter than our Sun, produce a photo-evaporation effect that inhibits planetary formation. [1]

Stars are composed mainly of the light elements hydrogen and helium. They also contain a small fraction of heavier elements such as iron, and this fraction is referred to as a star's metallicity . Stars of higher metallicity are much more likely to have planets, and the planets they have tend to be more massive than those of lower-metallicity stars. It has also been shown that stars with planets are more likely to be deficient in lithium. [1]

Life in the Universe and Where to find it

Temperature and composition

It is possible to estimate the temperature of an exoplanet based on the intensity of the light it receives from its parent star. For example, the planet OGLE-2005-BLG-390Lb is estimated to have a surface temperature of roughly -220°C (roughly 50 K). However, such estimates may be substantially in error because they depend on the planet's usually unknown albedo, and because factors such as the greenhouse effect may introduce unknown complications. A few planets have had their temperature measured by observing the variation in infrared radiation as the planet moves around in its orbit and is eclipsed by its parent star. For example, the planet HD 189733b has been found to have an average temperature of 1205±9 K (932±9°C) on its dayside and 973±33 K (700±33°C) on its nightside. [1]

If a planet is detectable by both the radial-velocity and the transit methods, then both its true mass and its radius can be found. The planet's density can then be calculated. Planets with low density are inferred to be composed mainly of hydrogen and helium, while planets of intermediate density are inferred to have water as a major constituent. A planet of high density is believed to be rocky, like Earth and the other terrestrial planets of the Solar System. [1]

Spectroscopic measurements can be used to study a transiting planet's atmospheric composition. Water vapor, sodium vapor, methane, and carbon dioxide have been detected in the atmospheres of various exoplanets in this way. The technique might conceivably discover atmospheric characteristics that suggest the presence of life on an exoplanet, but no such discovery has yet been made. [1]

Another line of information about exo-planetary atmospheres comes from observations of orbital phase functions. Extrasolar planets have phases similar to the phases of the Moon. By observing the exact variation of brightness with phase, astronomers can calculate particle sizes in the atmospheres of planets. [1]

Stellar light becomes polarized when it interacts with atmospheric molecules, which could be detected with a polarimeter. So far, one planet has been studied by polarimetry. [1]

Geoffrey W. Marcy (born September 29, 1954, St. Clair Shores, Michigan) is an American astronomer, who is currently Professor of Astronomy at the University of California, Berkeley, famous for discovering more extrasolar planets than anyone else, 70 out of the first 100 to be discovered, along with R. Paul Butler and Debra Fischer.

Marcy confirmed Michel Mayor and Didier Queloz's discovery of the first extrasolar planet orbiting a Sun-like star – 51 Pegasi b. Other achievements have included discovering the first multiple planet system around a star similar to our own (Upsilon Andromedae), the first transiting planet around another star (HD209458b), the first extrasolar planet orbiting beyond 5 AU (55 Cancri d), and the first Neptune-sized planets (Gliese 436b and 55 Cancri e).

Marcy also:

Developed Method of Precise Doppler Measurements (3 meter/sec)

Found Evidence that Solar System May be Peculiar (Circular vs Eccentric Orbits)

Discovered First System of Planets Around a Sun-Like Star (Upsilon And)

Discovered First Transiting Planet Around another Star (HD209458)

Discovered First Candidate Saturn-Mass Planets (HD46375, HD16141)

Discovered First Extrasolar Planet Orbiting Beyond 5 AU (55 Cancri d)

Co-Discovered First Neptune-Sized Planets: Gliese 436b and 55 Cancri e [1]

Geoff Marcy

NRAO 100 meter line searches

After Marcy's work and since the planet finder satellite Kepler has been in service, several radio telescopes, including the 100 meter GBT at NRAO in Greenbank, West Virginia, have been doing spectroscopic line searched of the star system harboring planets. This is really the first cogent attempt to find life (of any sort) on an extraterrestrial planet. The work will prove to be tedious but will ultimately create a data base of line emissions that will be considered for the indicators of a life process.

Example of Occultation

The following plot shows the occultation of a star by Pluto, done in June of 2006 at the Anglo-Australian Telescope in Australia. A 4.3 meter telescope was used with a special low light level digital camera. These data help scientists determine the temperature and pressure profiles of the atmosphere. In addition the presence of clouds and other atmospheric layer phenomena can be seen.

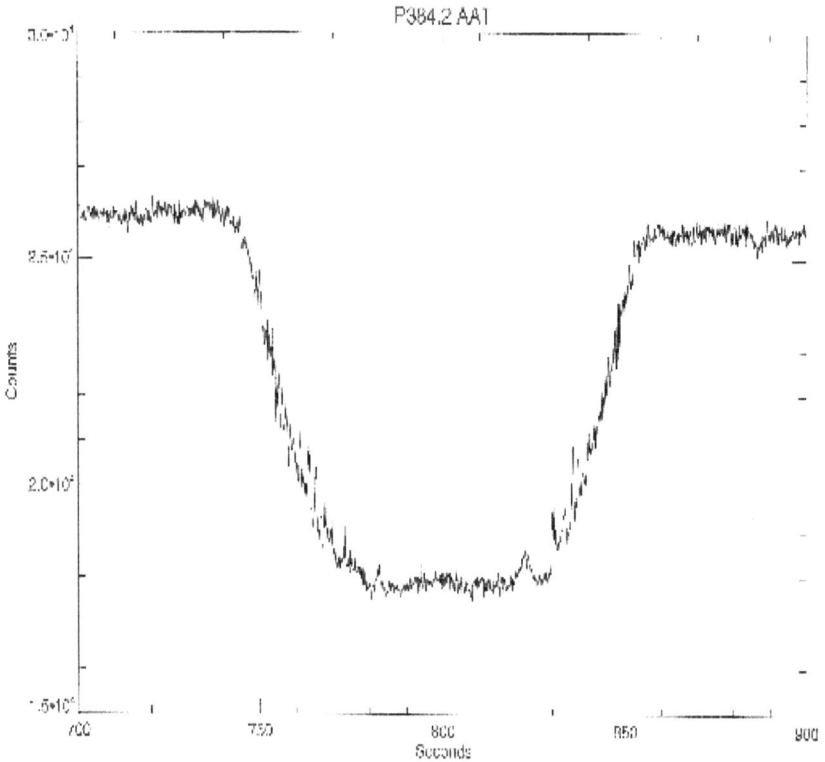

Occultation of Pluto showing characteristics of atmosphere, more on one side than the other. Spikes could indicate cloud layers

Conclusion

Thousands of new planets have been discovered using multiple methods.

Many more thousands of new planets will be discovered soon.

219

Chapter 10 – Life Here and There

Deep Sea Fish, which lives in up to 1000 atmospheres, 3 degrees C and no
Oxygen

Life in the Universe and Where to find it

Introduction

There is a wide spectrum of variables where life is concerned. It has been found at some of the deepest parts of the ocean, under enormous pressure and temperature. It has been found in the air, deep in the Earth, at the poles, on meteorites, and is suspected (for good reason) on several moons and possibly Mars. The general consensus in the biological community is that it is easy to start and hard to eliminate.

The spectrum of variables includes gravity, temperature, pressure, proximity to water, proximity to oxygen and the amount of localized harmful radiation.

Gravity

The boundaries of weight over the course of life on Earth has varied from 10 attograms for a simple virus to 172 metric tons for a Blue Whale. For plants, a Coastal Redwood can weigh up to 3,300 tons. Taking this a step further, evidence of bacterial life has been found on meteorites. If the Panspermia theory is correct, life can indeed exist in a weightless environment. There is obviously an abundance of life that is airborne such as seeds, pollen, mold, etc. However, as Mather once said, "Gravity makes it possible for us to be here."

Temperature

The boundaries of temperature that supports life is also quite vast. Life can exist in supercooled conditions at around 4 degrees Kelvin. Life has also been found in temperatures higher than 588 degrees Kelvin or over 600 degrees Fahrenheit. This high temperature life exists near deep ocean vents, were there is not only enormous temperature but also enormous pressures.

Pressure

The deep sea is also an extremely hostile environment, with pressures between 20 and 1,000 atmospheres (between 2 and 100 megapascals), life has been found at these pressure levels and at temperatures between 3 and 10 degrees Celsius, with a total lack of oxygen. 1,000 atmospheres corresponds to well over 14,000 psi.

Water

Only carbon based life requires water. There have been worms found very deep in South African Mines that live on chemicals that do not include water.

Air

There are many examples of life that do not require air. Anaerobic bacteria is a prime example at this. This type of bacteria is responsible for decomposition and putrefaction and requires no oxygen.

Food

As mentioned before, worms have been found that exist on chemicals alone and do not require food breakdown.

Radiation

Generally, life thrives in areas were there is a low amount of radiation, as in the so called "Goldilocks Orbits." However, several bacteria including two *Rubrobacter* species and the green alga *Dunaliella bardawil*, can endure high levels of radiation. *Deinococcus radiodurans*, on the other hand, can withstand up to 20 kiloGray (kGy) of gamma radiation and up to 1,000 joules per square meter of UV radiation. *D. radiodurans* can be exposed to levels of radiation that blow its genome into pieces only to have the organism repair its genome and be back to normal operations in a day. [1]

More on the "Goldilocks Orbits"

In astronomy, the habitable zone is the distance from a star where an Earth-like planet can maintain liquid water on its surface and Earth-like life. The habitable zone is the intersection of two regions that must both be favorable to life: one within a planetary, and the other within a galaxy. Planets and moons in these regions are the likeliest candidates to be habitable and thus capable of bearing extraterrestrial life similar to our own. [1]

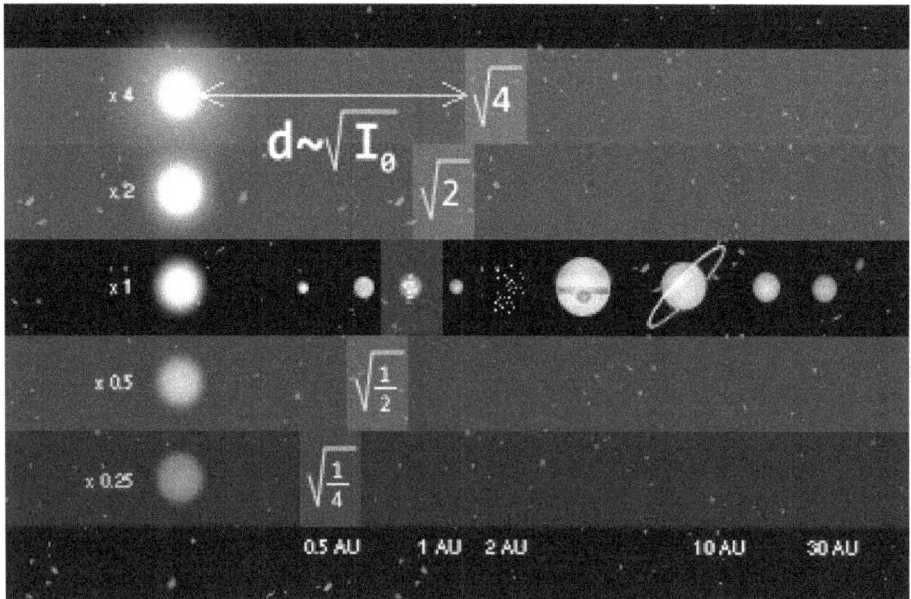

Diagram of Habitable Zones around a star.

223

Currently, astronomers believe that most stars evolve with a proto-planetary disk when "born." These are typically large amounts of dust and smaller rocks that coalesce into the sun as well as planets over the course of the birth. Eventually the planets clean out their orbits and a solar system is created. There are many variable with regards to how many planets and of what size. So far, there has been a significant amount of Jupiter and larger planets found. This make sense as they would logically be the first to be discovered as they perturb the stars light or motion more obviously than the smaller planets. There is also conjecture that indicates that a Jupiter like planet is good for a solar system as it cleans out many of the smaller asteroids, comets and other debris that could fall on other smaller planets.

Probably the best combination of planets in a solar system capable of harboring life will include smaller planets in the habitable zone with larger planets farther out. Much like our own solar system. Larger planets in "Earth's" orbit would have a gravity issue to contend with. Smaller planets farther away, for instance in Jupiter's orbit, would have the issue of not enough energy to keep the planet within the optimal temperature range.

Proto-planets in formation around a young star

There are also habitable zones in galaxies, where the amount of radiation from areas close into the center hubs, where large black holes are prevalent is dangerous.

Life in the Universe and Where to find it

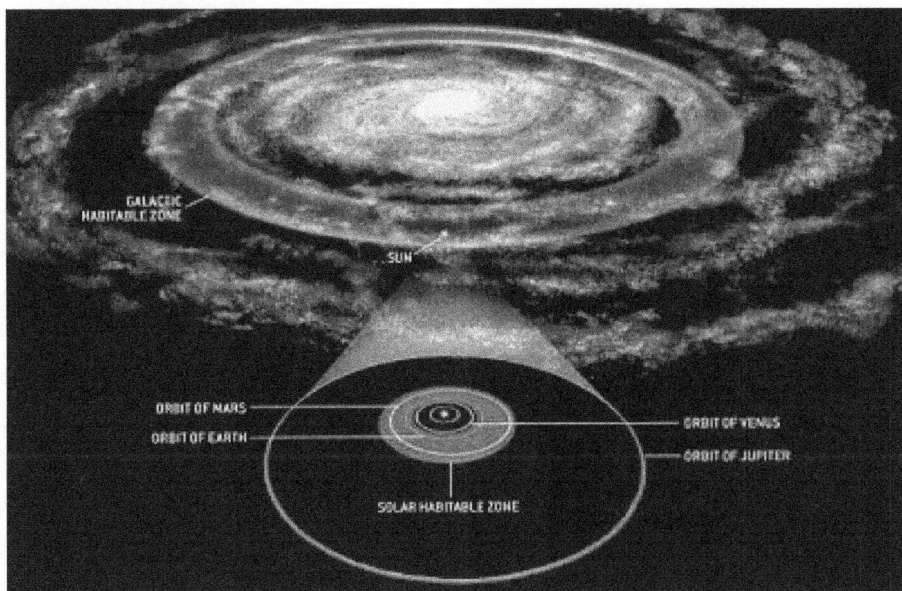

Optimal placement in the Milky Way

Biology of Life in the Universe

Extraterrestrial Life

Introduction

The two main theories on the origins of life seem to center on the *abiogenesis* idea of life from non life...and *panspermia*, where life came from the stars. The latter is getting support from biological markers in meteorites (usually from Antarctica, where contaminants are minimal). The latter also has religious interest where Johnson and others are attracted to the idea to

227

confirm Christian theology. The former idea is shown to have a viable start as in the Urey/Miller synthesis, but the bridge between this and self replicating cells is very complex process and a viable model has not yet been developed.

It is vitally important to remove religion from this argument as religion is not science and has the propensity of viewing the world through a "lens." Science by definition must be objective and open to any conclusion as long as it passes the test of time. The idea of *panspermia* however can still be valid without the religious content. This is due to the fact that life can be created even by *amniogenesis* and transported via meteorites, etc.

Abiogenesis has merit in experimentation starting with a very famous experiment by Miller and Urey in 1952 where they were able to combine water, the constituents of primitive Earth's atmophere and a spark (all available in primordial Earth) to produce Amino acids.

In natural science, *abiogenesis* or *biopoesis* is the study of how biological life arises from inorganic matter through natural processes, and the method by which life on Earth arose. Most amino acids, often called "the building blocks of life", can form via natural chemical reactions unrelated to life, as demonstrated in the Miller-Urey experiment and similar experiments that involved simulating some of the conditions of the early Earth in a laboratory. In all living things, these amino acids are organized into proteins, and the construction of these proteins is mediated by nucleic acids, that are themselves synthesized through biochemical pathways catalyzed by proteins. Which of these organic molecules first arose and how they formed the first life is the focus of abiogenesis.

Life in the Universe and Where to find it

In any theory of abiogenesis, two aspects of life have to be accounted for: replication and metabolism. The question of which came first gave rise to different types of theories. In the beginning, metabolism-first theories (Oparin coacervate) were proposed, and only later thinking gave rise to the modern, replication-first approach.

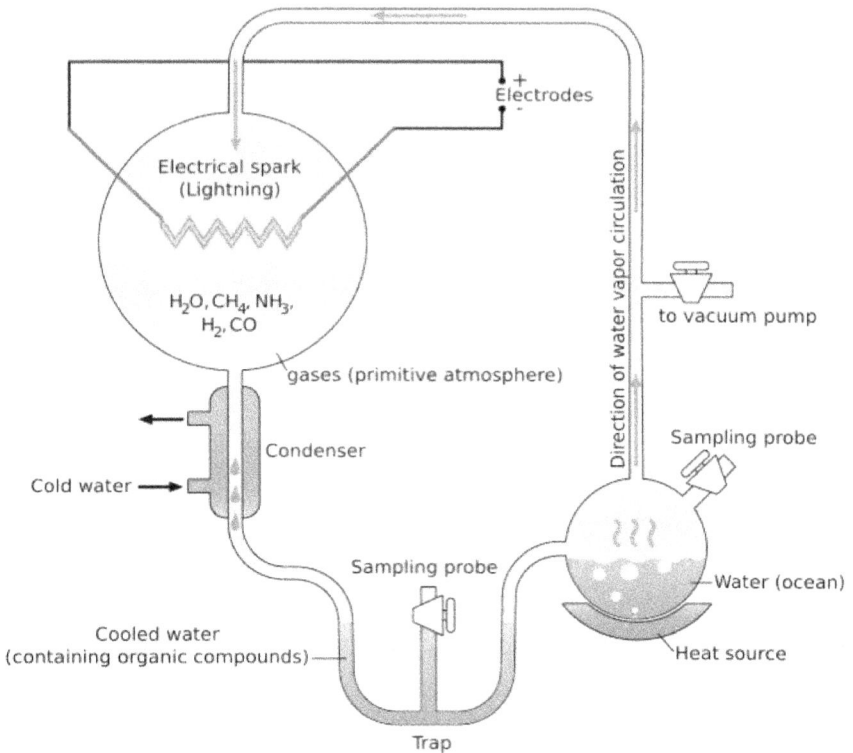

Synthesis of Amino Acids – The Miller and Urey Experiment

In modern, still somewhat limited understanding, the first living things on Earth are thought to be single cell prokaryotes (which lack a cell nucleus), perhaps evolved from protobionts (organic molecules surrounded by a membrane-like structure). The oldest ancient fossil microbe-like objects are dated to be 3.5 Ga (billion years old), approximately one billion years after the formation of the Earth itself. By 2.4 Ga, the ratio of stable isotopes of carbon, iron and sulfur shows the action of living things on inorganic minerals and sediments and molecular biomarkers indicate photosynthesis, demonstrating that life on Earth was widespread by this time.

The sequence of chemical events that led to the first nucleic acids is not known. Several hypotheses about early life have been proposed, most notably the iron-sulfur world theory (metabolism without genetics) and the RNA world hypothesis (RNA life-forms). [1]

After a series of pioneer explorations, the key experiment of Stanley Miller in the fifties, marked the origin of the prebiotic chemistry field using the demonstration that a reductive atmosphere over a water pond could generate amino acids, among other compounds (Miller, 1953). After this experiment and attempts to replicate the original composition of Archean atmosphere on Earth, laboratory simulations have led to the demonstration that many molecules of biochemical interest could be synthesized on water ponds under reducing, mildly reducing and even neutral atmospheric conditions. Also, the hydrothermal systems (in particular the deep sea hydrothermal vents) could provide the appropriate setting for the abiotic formation and accumulation of organics, providing precursors for the chemical evolution (i.e. Ferris, 1992; Simoneit 1995). [5]

The hydrothermal and volcanic environments also favor conditions

230

conducive to the *abiogenic* development of life due to the availability of sulphide and inorganic pyrophosphate, two reactants that are likely of importance to the final development of the first metabolism.

An analysis of prebiotic chemistry has shown us that a limited number of bulk material of biochemicals could be synthesized in a variety of conditions: amino acids, nucleobases and a set of carboxylic and hydroxycarboxylic acids.

Modern approaches (i.e. all the Günter Wächtershäuser papers) shows that a chemoautotrophic origin of life through a protometabolic, out of equilibrium, chemical system capable of carbon fixation, with the aid of surface of minerals and transition metal catalysts, could be a plausible alternative model for the "replication-first" models based on the autocatalytic polymerization of monomers and aggregation and replication of polymers (as RNA World). [1]

The Krebs Cycle

All cells, whether prokaryotic or eucaryotic, must produce energy to survive. The process behind the cellular energy production eluded scientists for years until 1937 when Hans A. Krebs proposed a specific metabolic pathway within the cells. The Krebs cycle, or citric acid cycle, was put forth to account for the oxidation of carbohydrates by animal tissues. Of course, later the acetyl derivative (a compound formed in fat degradation) was identical to the compound formed by the oxidative decarboxylation of pyruvate; thus proving that the Krebs cycle also serves for the oxidation of fats. Krebs' later work showed that the cycle is not restricted to animals, but rather is present in nearly all aerobic cells. Within the cell, the Krebs cycle takes place inside the mitochondria or "power plant" of the cell and provides the energy required for the organism to function.

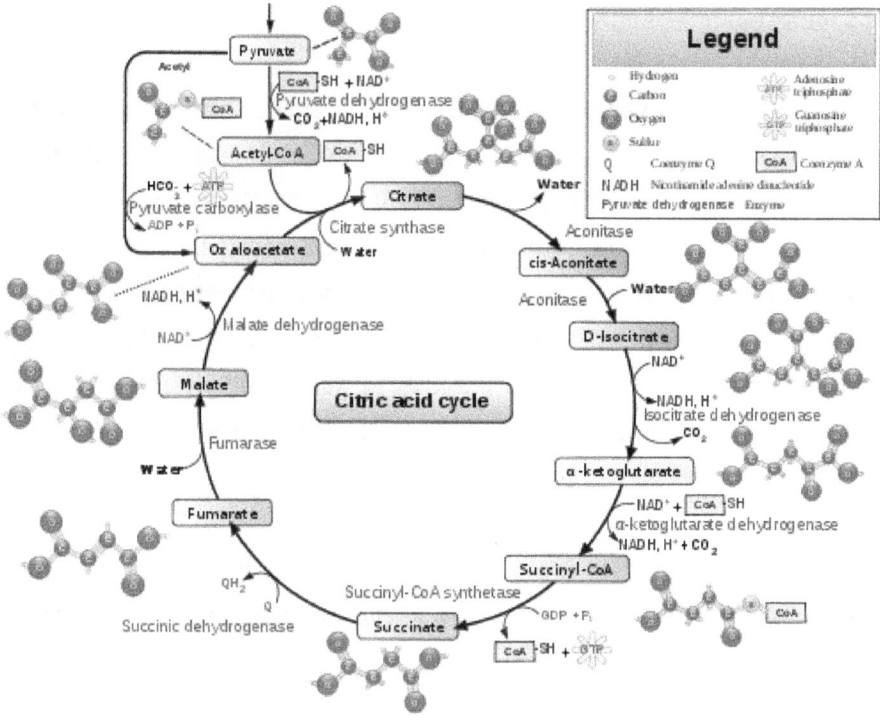

The Krebs or Citric Acid Cycle

A number of scientists are engaged in the study of origin of Krebs cycle or plausible variants, as reverse-Krebs cycle, and this has direct implications regarding the origin of life. This is due to two interesting facts: some hydroxycarboxylic acids implicated in the central metabolism are easily synthesized in abiotic conditions (Ruiz-Bermejo et al., 2007) and a chemical cycle could decrease the entropy and increase the complexity of a system, an interesting field that has their pioneer work in the Belousov-Zhabotinsky reaction. [1]

The Importance of Amino Acids

Life has been using a standard set of 20 amino acids to build proteins for more than 3 billion years," said Stephen J. Freeland of the NASA Astrobiology Institute at the University of Hawaii. "It's becoming increasingly clear that many other amino acids were plausible candidates, and although there's been speculation and even assumptions about what life was doing, there's been very little in the way of testable hypotheses." So Freeland and his University of Hawaii colleague Gayle K. Philip devised a test to try to learn if the 20 amino acids Earth's life uses were randomly chosen, or if they were the only possible ones that could have done the job.

Amino acids are molecules built primarily from carbon, hydrogen, oxygen, and nitrogen. They assemble in particular shapes and patterns to form larger molecules called proteins that carry out biological functions.

"Technically there is an infinite variety of amino acids," Freeland told Astrobiology Magazine. "Within that infinity there are lots more than the 20 that were available [when life originated on Earth] as far as we can tell." [1]

Testing the possibilities

The researchers defined a likely pool of candidate amino acids from which life drew its 20. They started with the amino acids that have been discovered within the so-called Murchison meteorite, a space rock that fell in

Murchison, Victoria in Australia in September 1969. The rock is thought to date from the early solar system, and to represent a sample of which compounds existed in the solar system and on Earth before life began.

Scientists have used computers to estimate the fundamental properties of the 20 amino acids life uses, such as size, charge and hydrophilicity, or the extent to which the molecules are attracted to water. These properties are important in the building of proteins. Freeland and Philip analyzed whether these properties could have been achieved with as much coverage and efficiency with other combinations of 20 amino acids. The researchers discovered that life seemingly did not choose its 20 building blocks randomly.

"We found that chance alone would be extremely unlikely to pick a set of amino acids that outperforms life's choice," Freeland said. [1]

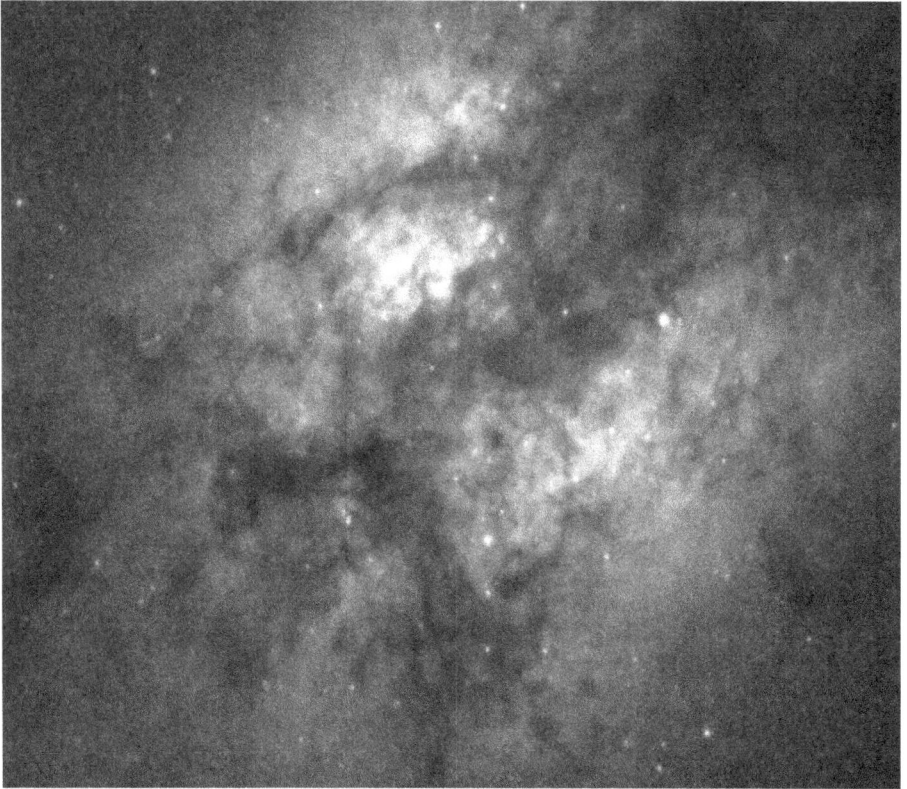

Astronomers have detected hydrogen cyanide and methanimine, two carbon compounds that can combine with water to make the amino acid glycine, in the Arp 220 galaxy shown above.

Natural selection

Researchers think early life on Earth probably used a version of natural selection to choose these amino acids. Some combinations of other amino acids were likely tried, but none proved quite as fit, so no other combinations ended up producing the numbers of successful offspring that the existing set achieved.

"Here we found a very simple test that begins to show us that life knew exactly what it was doing," Freeland said. "This is consistent with the idea that there was natural selection going on."

Getting at the question of why nature chose the 20 amino acids it did is experimentally difficult, said Aaron Burton, a NASA Postdoctoral Program Fellow who works as an astrochemist at NASA's Goddard Space Flight Center in Greenbelt, Md.

"Although a number of experiments have shown that unnatural amino acids can be incorporated into the genetic alphabet of organisms, it may never be possible to experimentally simulate sufficient evolutionary time periods to truly compare alternate amino acid alphabets," said Burton, who was not involved in the new study. "As a result, studies such as those presented by Philip and Freeland offer interesting insights and provide a framework for formulating hypotheses that can actually be tested in the lab." [1]

Amino acids in meteorites

Right now the race is on to directly find amino acids elsewhere in the solar system. Some hints that they abound have been found on meteorites that have landed on Earth from outer space, as well as from missions such as NASA's Stardust probe, which sampled the coma of comet Wild 2 in 2004.

"All signs are that amino acids are going to be found throughout the galaxy," Freeland said. "They are apparently obvious building blocks with which to construct life. What we're finding hints at a certain level of predictability in the way things turned out."

The question of life's amino acid toolbox is interesting not just in trying to trace the origin of the life on Earth, but in wondering whether life exists on other planets, and if so, what form it takes. Scientists are particularly curious about how a different set of amino acid building blocks would result in different characteristics in the life it creates.

"That is the biggest question of all," Freeland said. "We're trying to find a way to ask, if you change the set of amino acids with which we're building, what effect does that have on the proteins you can build. The most interesting thing is, nobody knows."

Anthropology

Ape from 2001: A Space Odyssey

Timeline of consciousness

If very hard to describe when humans or for that matter, other animals, became sentient or conscious of themselves. At the very least apes, monkeys and dolphins can problem solve and display thought processes that indicate sentience. For the purposes of this work, we will talk about human evolution to keep a coherent line of discussion. Consciousness most likely preceded writing by a significant degree and by some accounts came even before the Neanderthal era. Some anthropologists claim it started at least 100,000 years ago with evidence of the building of houses. Recently, 100,000 year old "makeup" was discovered; ocra in a seashell, possibly used for rituals. The cave paintings of Lascaux for instance are 17,300 year old.

Background

In lower forms of animal life, characteristics acquired by an individual cannot be passed on to its descendants.

During the beginnings of human society the knowledge and behavior of the individual is a distillation of minute advances occurring occasionally over thousands of generations. The specialization of function among individuals was no longer a strictly genetic trait, but a consequence of an increasingly planned and deliberate training to better serve the group as a whole. The character, skills, and behavioral patterns of an individual were as much influenced by upbringing and directed education as by inborn capabilities.

It is clear that further development of associative areas in the brain, together with better ability to utilize the knowledge residing in the internal model, would confer selective advantages. Brain size began to increase rapidly. Between the time of Australopithecus, some two million years ago, and the present day, the brain has increased in volume from about 500 cubic centimeters to 1300 or 1400 cubic centimeters.

The transmission of knowledge accumulated over many generations is known as cultural evolution. The now very sophisticated internal models of the external world allowed reasoning and the establishment of complex patterns of individual and group behavior based on that reasoning. The primitive instincts or "goals" of self-preservation, hunger, and reproductive drives still dominated, but the solutions to the daily problems of living were becoming more complex and successful. Man was becoming intelligent.

One of the characteristics of intelligence is that action can be taken to control the environment and hence to decrease its hazards. Early man developed a facility for the use of tools. At first, these were the simple instruments of stone and wood that litter dwelling sites. Gradual refinement brought the axe, the spear, and instruments to process game, to make clothes, and to manufacture further tools. Some half million years ago man learned to use fire. The materials of the environment were being manipulated to extend the physical capabilities of the body, and to supply energy for a variety of needs. Such activities could be described as central and largely unchallenged tenet of evolution.

As man evolved he began to develop many of the characteristics we classify as "human." The ability to retain crucial components of the internal model led to the beginnings of technology. the increasing complexity of short- and long-term memory. The model of the external world included a model of the individual himself. Consciousness and self-awareness are the words we use to describe this portion of the model. Individuals began to act in a concerted fashion, exhibiting group behavior. Codes of behavior were established, later to be refined into laws. Groups of humans were better able to defend themselves.

Around fifty to one hundred thousand years ago, the evolution of the human race had reached the point where Homo sapiens was emerging in something like his present form. Another species, Neanderthal man, with a somewhat different appearance, but with an equally large brain, existed until the Ice Ages, but then disappeared. The reasons are not clear. Two possibilities are an unfavorable geographic location at a time of intense climatic change, or competition with Homo Sapiens. Natural selection was still at work. [2]

Writing

If you include pictographs and petroglyphs, the communication of ideas via symbols began about 30,000 years ago. Formal writing, like that of the Sumerians appeared about 5,000 years ago. Evidence of ideas and numbers appeared in these writings and certainly, abstract thought was required to perform these functions.

Philosophizing

It is generally agreed that Thales was the first philosopher, but in reality philosophic thought predates him with the musings of the Delphic Oracles and others. This of course is from the perspective of Western Philosophy. Chinese, Sumerians and Egyptians no doubt had individuals that dwelled on heady matters, including life in the cosmos. It was here that cosmology had its origins.

Timeline of Religions

Religious thoughts have been around for many thousands of years and have evolved from animism to monotheism and some would say, beyond. Clearly, this is a human endeavor as no other species has exhibiting religious or spiritual worship. Of course the interesting question is whether or not extraterrestrials have spiritual feelings and if so, how are they expressed and in what context. The following categories of religions on Earth follow a chronological path and in all cases they still exist today.

Animism (100,000 BP-)

This religion refers to the belief that non-human entities are spiritual beings, or at least embody some kind of life-principle. It encompasses the beliefs that there is no separation between the spiritual and physical (or material) worlds, and souls or spirits exist, not only in humans, but also in all other animals, plants, rocks, natural phenomena such as thunder, geographic features such as mountains or rivers, or other entities of the natural environment. Animism may further attribute souls to abstract concepts such as words, true names, or metaphors in mythology. Animism is particularly widely found in the religions of indigenous peoples, perhaps most interestingly in Shinto and Sererism, and some forms of Hinduism, Buddhism, Pantheism, Christianity and Neopaganism.

Throughout European history, philosophers such as Aristotle and Thomas Aquinas, among others, contemplated the possibility that souls exist in animals, plants, and people; however, the currently accepted definition of animism was only developed in the 19th century by Sir Edward Tylor, who created it as "one of anthropology's earliest concepts, if not the first".

According to the anthropologist Tim Ingold, animism shares similarities to totemism but differs in its focus on individual spirit beings which help to perpetuate life, whereas totemism more typically holds that there is a primary source, such as the land itself or the ancestors, who provide the basis to life. Certain indigenous religious groups such as the Australian Aborigines are more typically totemic, whereas others like the Inuit are more typically animistic in their worldview. [1]

Polytheism (5,000 BP-)

This is the belief of multiple deities also usually assembled into a pantheon of gods and goddesses, along with their own mythologies and rituals. Polytheism was the typical form of religion during the Bronze Age and Iron Age, up to the Axial Age and the gradual development of monotheism or pantheism, and atheism. It is well documented in historical religions of Classical Antiquity, especially Greek polytheism and Roman polytheism, and after the decline of classical polytheism in tribal religions such as Germanic polytheism or Slavic polytheism. It persists into the modern period in traditions such as Hinduism, Buddhism, Shintoism, Chinese folk religion, etc., and it has been revived in currents of Neopaganism in the post-Christian West.

Polytheism is a type of theism. Within theism, it contrasts with monotheism, the belief in a singular God. Polytheists do not always worship all the gods equally, but can be Henotheists, specializing in the worship of one particular deity. Other polytheists can be Kathenotheists, worshipping different deities at different times. [1]

Monotheism (3,000 BP-)

This is the belief in the existence of one god as distinguished from polytheism, the belief in more than one god, and atheism, the absence of belief in any god. Monotheism is characteristic of the Baha'i Faith, Christianity, Druzism, Judaism, Islam, Sumaritanism, Sikhism and Zoroastrianism. It is difficult to delineate monotheism from beliefs such as pantheism and monism as in the Advaita traditions of Hinduism. Some scholars such as Wilhelm Schmidt argued for primeval monotheism: a monotheistic *Urreligion*, from which polytheistic religions developed.

Ostensibly monotheistic religions may still include concepts of a plurality of the divine; for example, the Trinity, in which God is one being in three eternal persons (the Father, the Son and the Holy Spirit). Additionally, most Christian churches teach Jesus to be two natures (divine and human), each possessing the full attributes of that nature, without mixture or intermingling of those attributes. This view is shared by the vast majority of Christians, with schisms regarding how the two natures are united.

Catholics and Eastern Orthodox venerate the saints, (among them Mary), as human beings who had remarkable qualities, lived their faith in God to the extreme and are believed to be capable of interceding in the process of salvation for others; however, Catholics do not worship (latria, properly translated as *adoration*, reserved for the Holy Trinity alone) them as gods, but instead offer dulia or hyperdulia to the saints and Mary respectively, properly translated as *veneration* or *to give homage*.

The concept of monotheism in Islam and Judaism rejects this distinction. Other forms of monotheism includes unitarianism and deism. [1]

Pantheism(2,500 BP-)

Pantheism is the view that the Universe, Nature and God (or divinity) are identical. Pantheists thus do not believe in a personal, anthropomorphic or creator god. The word derives from the Greek (pan) meaning "all" and the Greek (theos) meaning "God". As such, Pantheism denotes the idea that "God" is best seen as a process of relating to the Universe. Although there are divergences within Pantheism, the central ideas found in almost all versions are the Cosmos as an all-encompassing unity and the sacredness of Nature. [1]

Atheism (200 BP-)

Atheism is, in a broad sense, the rejection of belief in the existence of deities. In a narrower sense, atheism is specifically the position that there are no deities. Most inclusively, atheism is simply the absence of belief that any deities exist. Atheism is contrasted with theism which in its most general form is the belief that at least one deity exists.

The term *atheism* originated from the Greek *atheos*, meaning "without god", which was applied with a negative connotation to those thought to reject the gods worshipped by the larger society. With the spread of free thought, skeptical inquiry, and subsequent increase in criticism of religion, application of the term narrowed in scope. The first individuals to identify themselves as "atheist" appeared in the 18th century.

Atheists tend to be skeptical of supernatural claims, citing a lack of empirical evidence. Atheists have offered various rationales for not believing in any deity. These include the problem of evil, the argument from inconsistent revelations, and the argument from non-belief. Other arguments for atheism range from the philosophical to the social to the historical. Although some atheists have adopted secular philosophies, there is no one ideology or set of behaviors to which all atheists adhere.

In Western culture, some atheists are frequently assumed to be irreligious, although other atheists are spiritual. Moreover, atheism also figures in certain religious and spiritual belief systems, such as Jainism, Buddhism, Hinduism, and Neopagan Movements such as Wicca. Jainism and some forms of Buddhism do not advocate belief in gods, whereas Hinduism holds atheism to be valid, but difficult to follow spiritually. [1]

Paleontology

As paleontologist, archeologists and anthropologists uncover more and more artifacts from our ancestors, we are getting a much better picture of our history over the ages. These discoveries dove tail with geology and climatology to produce a time line of the evolution of life on Earth.

After the solar system was formed from a stellar disk and the planets coalesced, what drove the timeline of life was the environment. Even in the volcanic era of Pangea, life became abundant. For millions of years after the first evidence of life, oxygen was not available. After this point and interestingly, after significant amounts of meteorite bombardment, life flourished and diversified. Species begat species over many billions of years and as of now, 99% of all species that have graced the face of the Earth are

Life in the Universe and Where to find it

extinct. The following table shows the timeline of life on Earth:

Event	When it Happened on Earth (Myrs)	How long It Took to Complete (Myrs)	Possible Minimum Time (Myrs)
Origin of life	3800–3500	<500	10
Oxygen photosynthesis	<3500	<500	Negligible
Oxygen environments	2500	1000	100
Tissue multicellularity	550	2000	Negligible
Development of animals	510	5	5
Land ecosystems	400	100	5
Animal intelligence	250	150	5
Human intelligence	3	3	3

Researchers have several methods to determine the age of fossils. Some use the placement of the artifact in the known geological strata as an indicator of age. There are several other methods like Carbon 14 and Florine dating that use the natural decay of the compound once the animal or plant has died. Using multiple methods, scientists have been able to determine and confirm the timeline of life on Earth. As new methods appear such as Potassium 40 and Uranium 235 decay analysis, the confirmations continues.

A graphic depiction of life on Earth is presented below. It shows the rapid development of species in the relatively recent past. This rapid development includes for instance the evolution of mammals, humans in particular and of intelligence.

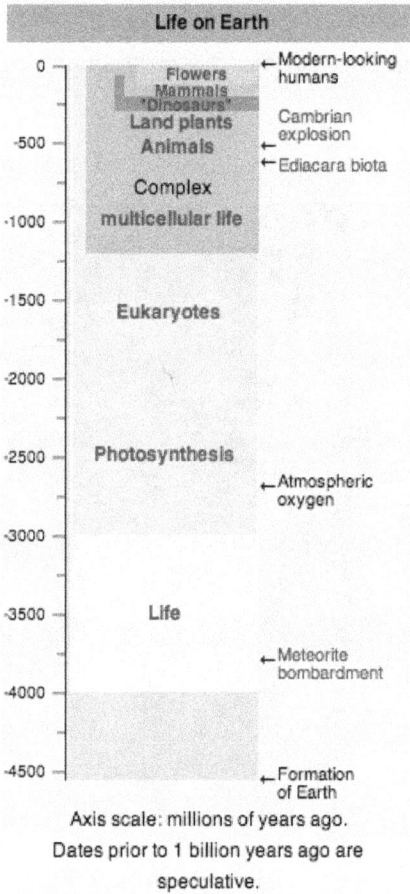

Life on Earth

```
   0 ─┤           Flowers          ← Modern-looking
                  Mammals             humans
                 "Dinosaurs"
                 Land plants        Cambrian
  -500 ─┤        Animals            explosion
                                    ←
                                    ← Ediacara biota
                 Complex
 -1000 ─┤      multicellular life


 -1500 ─┤        Eukaryotes


 -2000 ─┤


 -2500 ─┤      Photosynthesis
                                   ← Atmospheric
                                      oxygen
 -3000 ─┤


 -3500 ─┤          Life

                                   ← Meteorite
                                      bombardment
 -4000 ─┤


 -4500 ─┤                          ← Formation
                                      of Earth
```

Axis scale: millions of years ago.
Dates prior to 1 billion years ago are
speculative.

One needs to keep in mind that there was a huge die off of dinosaurs and other species approximately 65 Million years ago. Other catastrophic events have been discovered as well in history. In all cases, life finds a way to continue, albeit on a completely new path. Its interesting to speculate what the Earth would be like if for instance the event 65 Million years ago did not happen, or if there had been more events like it.

The human time line is also interesting where the size of the brain has

increased remarkably over the last few million years. Notice in the chart below
the relationship of brain size, teeth and bipedalism are related over the last 8
million years.

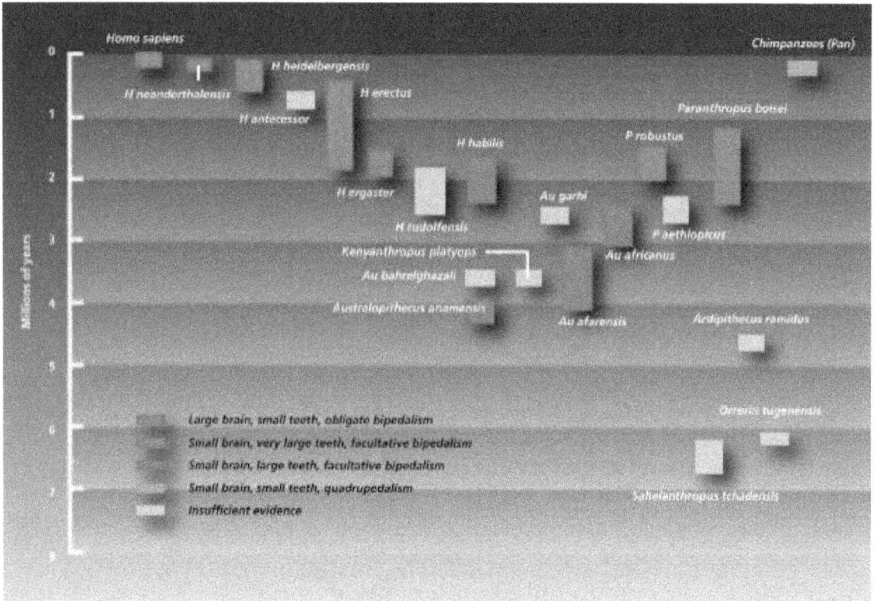

Evolution of Hominids over time

Development of the Hominid Brain

The graph below is modified from Dean Falk, *Hominid Brain Evolution: Looks Can Be Deceiving* in **Science** (1998) 280:1714. Additions to data, in numerals, are footnoted below.

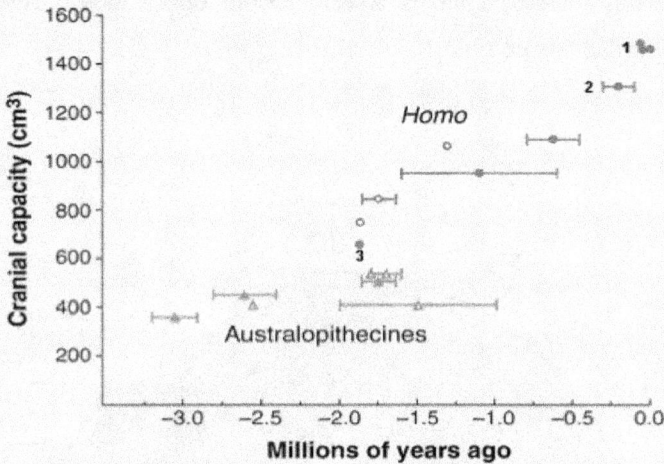

1. Homo neanderthalensis exceeds 1450 cc.
2. Modern Homo sapiens averages ~1350 cc.
3. Vekua A. et al. *A New Skull of Early* Homo *from Dmanisi, Georgia*. **Science** (2002), 297: 85-88. (Homo erectus/ergaster) ~ 1.75 mybp. Endocranial volume ~600 cc.

Brain size over time

250

Timeline of life – Absolute and Relative

Viewing evolution as it relates to geological time offers insight into how long certain events take. Life in simple form has been around for billions of years, whereas intelligent life is a very recent occurrence. The following graphic illustrates this point:

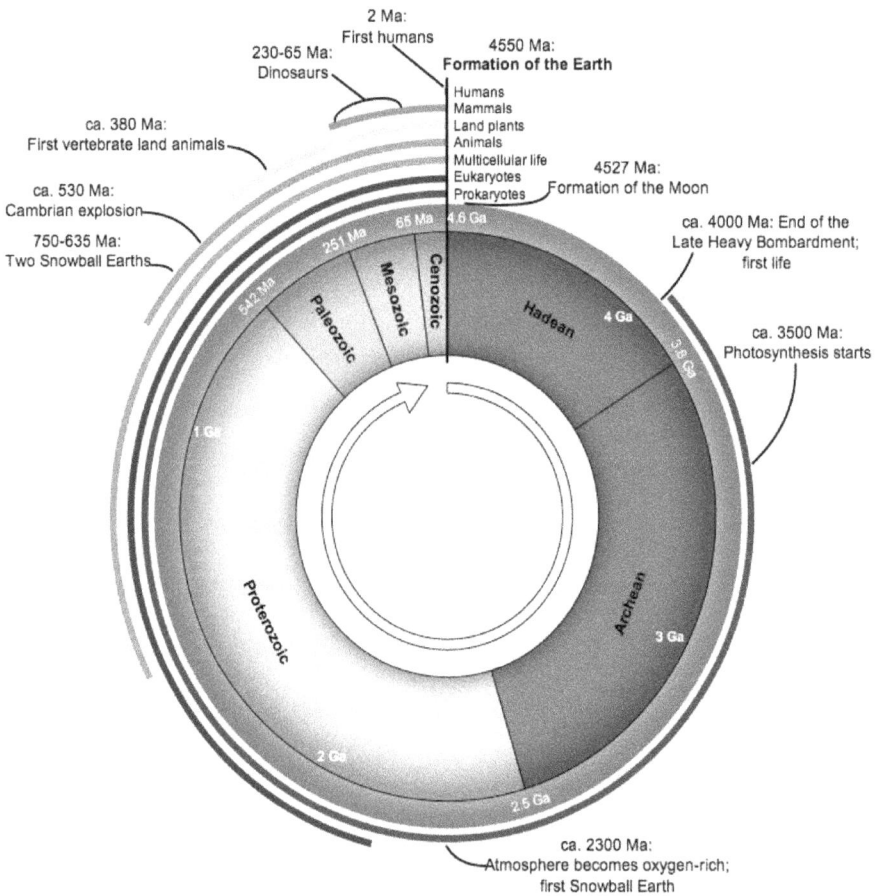

Geologic vs. Biological Time Line

251

Life in the Universe and Where to find it

A truncated list of the development of life is as follows:

First detected life forms on Earth

Simple cells (prokaryotes) 3.8 billion years ago

First Plants

photosynthesis 3 billion

First complex cells

eukaryotes 2 billion

Multicellular life

1 billion

First simple animals

600 Million

First arthropods

(ancestors of insects, arachnids and

crustaceans) 570 Million

Complex Animals

550 Million

First Fish (proto amphibians)

500 Million

Land Plants

475 Million

Insects and seeds

400 Million

Amphibians

360 Million

Reptiles

300 Million

Mammals

200 Million

Birds

150 Million

Flowers

130 Million

Non-avian dinosaurs died out

65 Million

Genus Homo appearance

2.5 Million

Common Human appearance

200,000

Neanderthals

25,000

History of Human Beings

Focusing on just human beings the following chart shows the species and time period our ancestors lived:

SPECIES	TIME PERIOD
Ardipithicus ramidus	5 to 4 million years ago
Australopithecus anamensis	4.2 to 3.9 million years ago
Australopithecus afarensis	4 to 2.7 million years ago
Australopithecus africanus	3 to 2 million years ago
Australopithecus robustus	2.2 to 1.6 million years ago
Homo habilis	2.2 to 1.6 million years ago
Homo erectus	2.0 to 0.4 million years ago

Homo sapiens archaic	400 to 200 thousand years ago
Homo sapiens neandertalensis	200 to 30 thousand years ago
Homo sapiens sapiens	200 thousand years ago to present

Brain size has increased dramatically as can be seen in the following graph:

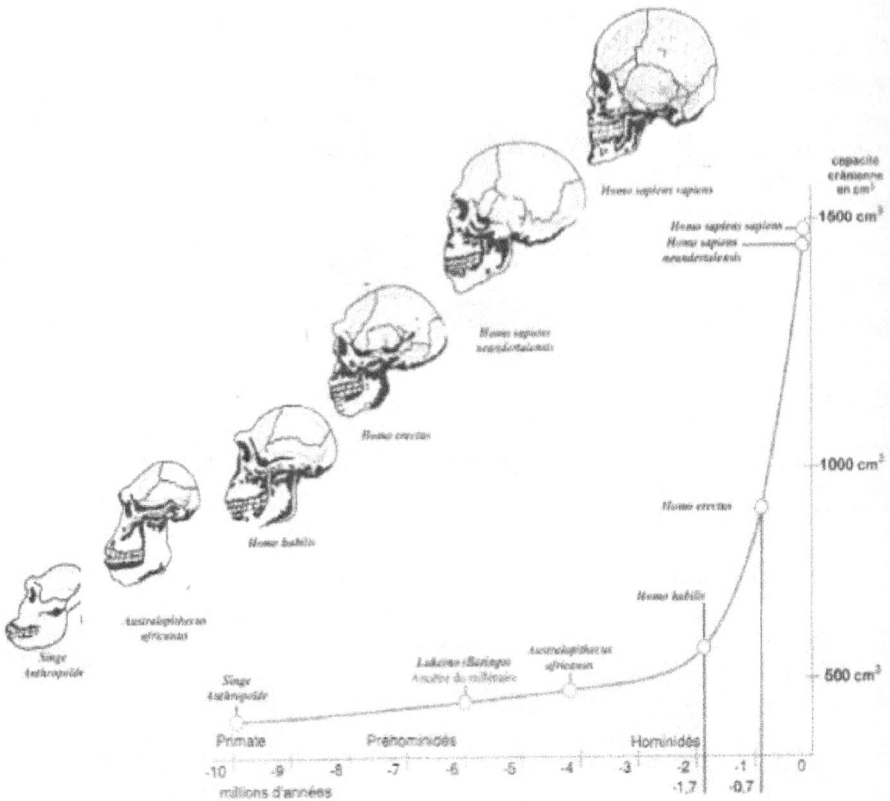

Size of brains over the last 10 Million years

Geology

The geologic time line illustrates the different geologic periods that are differentiated by changes in temperature, ice ages, vulcanism and biological evolution.

Geologic Periods

What is interesting about the advent of life on Earth is how quickly it started after the solar system had formed. As soon as the temperature and poisonous atmospheric gases became tolerable, the first simple single cell entities appeared.

The Advent of Technology

The importance of technology is the subject of many discussions, papers and books. From a philosophical point of view, it can define us and displace our humanity. This latter concept is certainly true when military personnel conduct war operations from many miles away from the battle theatre using robotic aircraft and other remotely controlled vehicles.

For the purposes of our discussions however its important to understand the usefulness of the tools technology has created. It is certainly by these inventions that spacecraft will explore the stars and before that, discover planets with life on them. Inventions have been around for just a very small portion of intelligent life's existence. However the pace of inventions is increasing rapidly as can be seen by the number of patents applied for recently. There has been an exponential growth in these applications as well as Intellectual Property, which is the protection of inventions, designs, written materials and software.

No doubt technology is here to stay and is becoming more and more integrated into society and personal lives. The rapidity of learning it's nuances by young people indicates that it is in a way natural, allowing us to move quicker, use our minds better and produce more. It will be our vehicle for the exploration of space.

The following list shows how much has been achieved over the last 4,400 years:

2400 BC   The abacus, the first known calculator, invented in Babylonia

300 BC   Pingala invents the binary number system

87 BC   Antikythera Mechanism invented in Rhodes to track movement of the stars

724   Liang Ling-Can invents the first fully mechanical clock

1041   Movable type printing press invented by Bi Sheng

1280   Eyeglasses were invented

1350   Suspension bridges built in Peru

1450   Alphabetic, movable type printing press invented by Johann Gutenberg

1500   Ball bearing invented by Leonardo Da Vinci together with flying machines, including a helicopter, the first mechanical calculator and one of the first programmable robots

Life in the Universe and Where to find it

1510	Pocket watch invented by Peter Henlein
1576	Ironclad warship invented by Oda Nobunaga
1581	Pendulum invented by Galileo Galilei
1593	Thermometer invented by Galileo Galilei
1608	Telescope invented by Hans Lippershey
1609	Microscope invented by Galileo Galilei
1642	Adding machine invented by Blaise Pascal
1643	Barometer invented by Evangelista Torricelli
1698	Steam engine invented by Thomas Savery
1671	Gottfried Leibniz is known as one of the founding fathers of calculus (along with Newton)
1705	Steam piston engine invented by Thomas Newcomen
1708	Jethro Tull invents mechanical (seed) sower
1710	Thermometer invented by Rene Antoine Ferchault de Reaumur

1733	John Kay invents flying shuttle.
1742	Franklin stove invented by Benjamin Franklin
1752	Lightning rod invented by Benjamin Franklin
1767	Spinning Jenny invented by James Hargreaves
1769	Steam engine invented by James Watt
1774	Priestly isolates oxygen
1779	First steam powered mills automate the weaving process.
1781	William Herschel discovers the planet Uranus.
1783	Hot air balloon invented by Montgolfier brothers
1791	Steamboat invented
1645	Vacuum pump invented by Otto von Guericke
1657	Pendulum clock invented by Christiaan Huygens
1679	Pressure cooker invented by Denis Papin
1687	Newton, Principia: Newton's physics formed the foundation of

modern science

1793   Eli Whitney develops the cotton gin

1798   Vaccination invented by Edward Jenner

1799   Humphrey Davy discovers nitrous oxide (laughing gas)

            Oliver Evans invents the conveyer belt

1804   Locomotive invented by Richard Trevithick

1814   Steam Locomotive (Blucher) invented by George Stephenson

1816   Miner's safety lamp invented by Humphry Davy

1816   Stethoscope invented by Rene Theophile Hyacinthe Laennec

1820   The Arithmometer was the first mass-produced calculator invented by Charles Xavier Thomas de Colmar

1821   Faraday demonstrates the principle of the electric motor.

1822   Charles Babbage designs his first mechanical computer

1821   Electric motor invented by Michael Faraday

1826   Photography invented by Joseph Nicephore Niepce

1830	Lawn mower invented by Edwin Beard Budding
1831	Von Liebig discovers chloroform
	Faraday discovers electro-magnetic current, making possible generators and electric engines.
1834	Braille invented by Louis Braille
1834	Refrigerator invented by Jacob Perkins
1834	Combine harvester invented by Hiram Moore
1835	Morse code invented by Samuel Morse
	Revolver invented by Samuel Colt
1838	Electric telegraph invented by Charles Wheatstone (also Samuel Morse)
	Daguerre perfects the Daguerrotype.
1839	Vulcanization of rubber invented by Charles Goodyear
1842	Anaesthesia invented by Crawford Long
1843	Typewriter invented by Charles Thurber
1846	Sewing machine invented by Elias Howe

1846   Rotary printing press invented by Richard M. Hoe

            Pneumatic tire patented

            Elias Howe invents the sewing machine

1849   Safety pin invented by Walter Hunt

            Monier develops reinforced concrete

1859   Charles Darwin publishes The Origin of Species

            Etienne Lenoir demonstrates the first successful gasoline
            engine

1862   Revolving machine gun invented by Richard J. Gatling

            Isaac Singer commercializes the sewing machine

1862   Mechanical submarine invented by Narcís Monturiol i Estarriol

1866   Dynamite invented by Alfred Nobel

            Lister demonstrates the use of carbolic antiseptic

1869   Mendeleev produces the Periodic Table

1870   Stock ticker invented by Thomas Alva Edison

1873	Christopher Sholes invents the Remington typewriter
	James Clerk Maxwell states the laws of electro-magnetic radiation
1876	Gasoline carburettor invented by Daimler
1877	Phonograph invented by Thomas Alva Edison
1877	Microphone invented by Emile Berliner
1878	Cathode ray tube invented by William Crookes
	Edison invents the incandescent lamp.
1880	Photophone invented by Alexander Graham Bell
1883	First skyscraper built in Chicago (ten stories)
	Maxim invents the machine gun
1885	Motor cycle invented by Gottlieb Daimler and Wilhelm Maybach
	Benz develops first automobile to run on internal- combustion engine
1888	Hertz produces radio waves.

Life in the Universe and Where to find it

1888   Eiffel Tower is bult

1891   Zipper invented by Whitcomb L. Judson

            Rudolf Diesel invents diesel

1893   Wireless communication invented by Nikola Tesla

1895   Diesel engine invented by Rudolf Diesel

            Radio signals were invented by Guglielmo Marconi

            Auguste and Louis Lumiere develop Cinematograph

            Roentgen discovers X-rays

1898   Remote control invented by Nikola Tesla

1900   Planck develops quantum theory

            First Zeppelin is built

1901   Vacuum cleaner invented by Hubert Booth

1903   Powered airplane invented by Wilbur Wright and Orville Wright

1905   Einstein writes the Theory of Relativity.

1907   Color photography invented by Auguste and Louis Lumiere

Life in the Universe and Where to find it

Helicopter invented by Paul Cornu

Radio amplifier invented by Lee DeForest

1908   Henry Ford mass-produces the Model T

1909   Bakelite invented by Leo Baekeland

1919   London to Paris air service begins

James Smathers develops the first electric typewriter

1923   Sound film invented by Lee DeForest

Television Electronic invented by Philo Farnsworth

1924   Electro Mechanical television system invented by John Logie
Baird

1926   Robert Goddard experiments with liquid-fueled rockets.

1928   Antibiotics, penicillin invented by Alexander Fleming

1931   Iconoscope invented by Vladimir Zworykin

1937   Jet engine invented by Frank Whittle and Hans von Ohain

Alan Turing develops the concept of a theoretical computing
machine

Life in the Universe and Where to find it

1938	Ballpoint pen invented by Laszlo Biro
1943	Enigma: Adolf Hitler uses the Enigma encryption machine
	Colossus: Alan Turing develops the the code-breaking machine Colossus
	Aqua-Lung invented by Jacques-Yves Cousteau and Emile Gagnan
1945	The atomic bomb
1946	Microwave oven invented by Percy Spencer
1951	Nuclear power reactor invented by Walter Zinn
1956	Optical fiber invented by Basil Hirschowitz, C. Wilbur Peters, and Lawrence E. Curtiss
	Videocassette recorder invented by Ampex
1957	Sputnik I and Sputnik II: Sputnik I and Sputnik II are launched by the Russians
1958	Silicon chip: The first integrated circuit, or silicon chip, is produced by the US Jack Kilby & Robert Noyce
1960	Laser invented by Theodore Harold Maiman

Life in the Universe and Where to find it

1961	Uri Gagarin is the first man in space
	Optical disc invented by David Paul Gregg
1963	Computer mouse invented by Douglas Engelbart
1967	Automatic Teller Machine (ATM) invented by John Shepherd-Barron
	Hypertext invented by Andries van Dam and Ted Nelson
1968	Video game console invented by Ralph H. Baer
1969	The moon landing - Neil Armstrong sets foot on the moon
1971	E-mail invented by Ray Tomlinson
	Liquid Crystal Display invented by James Fergason
	Pocket calculator invented by Sharp Corporation
	Floppy Disk invented by David Noble with IBM
1973	Ethernet invented by Bob Metcalfe and David Boggs
	Personal computer invented by Xerox PARC
1983	Camcorder invented by Sony

1990   World Wide Web invented by Tim Berners-Lee

2001   Digital satellite radio

There are now many thousands of inventions, with many more being created every day. Soon, human beings will be born in space, and may never know life on a planet. They will be surrounded by technology and fully dependent on it. One wonders how another planet's life forms will react to this person as they are just as stunned walking off of the starship and viewing a natural world, as the indigenous population will be seeing the stranger.

Conclusion

Our system worked.

Life can be produced easily by either natural chemical progressions and/or propagation from space.

Life is resilient and can exist in very harsh environments.

Life is very adaptive.

The progress of sentient intellectual advancements is unique.

Human progress in general is unique and rapidly accelerating.

Life in all forms on Earth is unique.

Chapter 11 - Timelines

The following pages show the timelines of more technical achievements and make clear how human beings have advanced in the very recent past. Keep in mind for instance Moore's Law, where computing power doubles every two years. Since the law was proclaimed in 1965, it has been uncannily accurate and is the basis for predicting future growth in the semi-conductor industry. Moore noticed that since 1958, the number of transistors had doubled every year until his paper of 1965. He then postulated that a trend of doubling the number of transistors every two years would continue for at least 10 more years. It has of course continued to do so well past his predictions. The development of technology has certainly been impressive in the computer field as well as others, as is illustrated in this chapter.

Aviation

First Glider Flight, Mar 17, 1883 - John J Montgomery began the first of a series of glider flights at Otay Mesa (near El Cajon) CA.

First Airplane Factory, 1900 - Carl Dryden Browne started a commercial airplane factory in Freedom KS and built a model, but was unable to perfect his aircraft. The factory closed in 1902.

First Airport - Dec 1903 - Contestable in light of many previous open fields and hillsides that were used for glider and balloons flights; however, since Kitty Hawk had a hangar and a workshop, as well as its own weather station (courtesy of the Coast Guard), and it was the birthing grounds for the Air Age, we feel it is quite proper to regard it as the world's very first dedicated airport.

First Man-controlled powered flight - Dec 17, 1903 - Orville Wright, at Kitty Hawk NC, in a 12-second flight of 120 feet. (Wilbur Wright had tried three days earlier, and failed to get airborne; however, he made the fourth flight this historic day (after three previous short hops) for 852' in 59 seconds.)

First Fully controllable and maneuverable flight - June 23, 1905 - First flight of the Wright Flyer #3 at Huffman Prairie, Dayton OH. By October this aircraft was fully controllable, able to turn and bank, and remain aloft for up to 38 minutes.

First Night flight - June, 1910 - Charles W Hamilton, over Knoxville TN. A similar claim is made for Walter Brookins, also in 1910, for a night flight in a Wright at Montgomery AL.

First Airplane flown from a ship - Nov 14, 1910 - Eugene Ely, in Curtiss Albany Flyer, from an 83-foot platform on battleship USS *Birmingham*.

First Water takeoff and landing - Jan 26, 1911 - Glenn Curtiss, in his Hydro, at San Diego. (Henri Fabre had had made a successful a water take-off in his Hydroavion monoplane on 3/28/10, at Martigues, France, but could not alight on water because of fragility of his airfoil pontoons and had to land on shore instead.)

First Gyroscopically controlled autopilot, June, 1914 - Lawrence Sperry, son of inventor Elmer A Sperry, flew a Curtiss flying boat fitted with four gyroscopes over the Seine River in France while standing in the cockpit with his hands clearly off the controls, and his mechanic standing on the lower wing. Despite gusty winds, the ship maintained longitudinal stability and won a 50,000-franc first prize.

First Transatlantic flight, May 16-31, 1919 - USN LtCdr Albert C Read and pilot Lt Walter Hinton, in Curtiss NC-4, one of a flight of three, from Long Island to Plymouth, England, via Newfoundland, the Azores, and Lisbon, 4,514 miles in 53h:58m.

First Lifting body, 1920 - Patent application for a winged passenger-carrying airfoil-fuselage by Vincent J Burnelli was filed Jan 6, 1921 but, for unknown reasons, issue (#1,758,498) was delayed until May 13, 1930. His Remington-Burnelli RB-1, however, flew successfully in 1920 to prove feasibilty of the design, which carried on in many subsequent models and variants into the 1960s.

First All-metal airplane, June 20, 1920 - Gallaudet CO-1.

First In-flight plane-to-plane refueling, June 26, 1923 - Capt L H Smith and Lt J P Richter, in an Army de Havilland DH-4B over Rockwell Field, San Diego. They also set a distance record of 3,293 miles covered in the flight.

First Flight on instruments, Mar 7, 1924 - USAS Lts E H Barksdale and B Jones, for 575 miles in a flight from Dayton OH to Mitchel Field NY in a DH-4B.

First Transatlantic non-stop solo flight, May 20-21, 1927 - Charles A Lindbergh, in Ryan NYP *Spirit of St Louis,* from Long Island to Paris, in

33h:32m (?>33h:39m). Although 91 persons in 13 separate flights had crossed the Atlantic before him, he flew directly between two great world cities and did it alone. Incidental to that flight, he also set several new intercity speed records in flying the Ryan from San Diego to New York.

First Woman to cross the Atlantic by air, June 17-18, 1928 - Amelia Earhart, from Newfoundland to Wales, as a passenger in the Wilmer Stultz-Louis Gordon Fokker C-2 *Friendship.*

First Transatlantic scheduled airline, May 20, 1939 - Pan-American Airways, from New York to Portugal and France, in a Boeing 314 *Yankee Clipper,* at first only air mail. Pathfinder flight from Baltimore on Mar 26, 1939 carried 21 passengers. While the *Graf Zeppelin* did commercially connect Europe and the Americas in the mid-1930s, it was the airplane that would represent the means for fast, scheduled air service.

First Jet airplane, Oct 1, 1942 - Bell XP-59A, piloted by Robert Stanley, at Muroc AFB CA. What jetplane flew first? Germany's Heinkel He.178 flew on 8/27/39, the Italian Campini-Caproni CC-2 (not a true jet, but more of a turbojet, actually a thermojet) on 8/27/40, and the British Gloster E.28/39 on 5/15/41.

First Airplane to break the sound barrier, Oct 14, 1947 - Bell XS-1, flown to Mach 1.06 (700mph) by USAF Capt Charles E Yeager at Muroc AFB CA.

First Airplane to exceed Mach 2, Nov 20, 1953 - Douglas D-558-2, flown by A Scott Crossfield at Edwards AFB CA. The plane—designed for a maximum speed of Mach 1.6—was released from a B-29 mother ship at 32,000' and climbed to 72,000' to enter a shallow dive, reaching Mach 2.005.

First Airplane to exceed Mach 3, May, 1965 - Lockheed A-12/SR-71.

First aircraft to exceed Mach 6, The North American X-15 rocket powered aircraft/spaceplane was part of the X-series of experimental aircraft, initiated with the Bell X-1, that were made for the U.S. Air Force, NASA and the U.S. Navy. The X-15 set speed and altitude records in the early 1960s, reaching the edge of outer space and returning with valuable data used in aircraft and spacecraft design. As of 2011, it holds the official world record for the fastest speed ever reached by a manned rocket-powered aircraft.

X-15 Performance

Maximum speed:	Mach 6.72 (4,520 mph, 7,274 km/h)
Range:	280 mi (450 km)
Service Ceiling:	67 mi (108 km, 354,330 ft)
Rate of Climb:	60,000 ft/min (18,288 m/min)
Wind loading:	170 lb/ft^2 (829 kg/m^2)
Thrust/Weight ratio	2.07

First Commercial Supersonic Aircraft, 1969, Aérospatiale-BAC Concorde was a turbo-jet powered supersonic passenger airliner or supersonic transport (SST). It was a product of an Anglo-French government treaty, combining the manufacturing efforts of Aerospatiale and the British

Aircraft Corporation. First flown in 1969, Concorde entered service in 1976 and continued commercial flights for 27 years. While commercial jets take eight hours to fly from New York to Paris, the average supersonic flight time on the transatlantic routes was just under 3.5 hours. Concorde had a maximum cruise altitude of 18,300 metres (60,039 ft) and an average cruise speed of Mach 2.02, about 1155 knots (2140 km/h or 1334 mph), more than twice the speed of conventional aircraft.

First Jumbo Jet, The Boeing 747 is a wide-body commercial airliner and cargo transport, often referred to by its original nickname, *Jumbo Jet*, or *Queen of the Skies*. It is among the world's most recognizable aircraft,[4] and was the first wide-body ever produced. Manufactured by Boeing's Commercial Airplane unit in the United States, the original version of the 747 was two and a half times the size of the Boeing 707, one of the common large commercial aircraft of the 1960s. First flown commercially in 1970, the 747 held the passenger capacity record for 37 years.

First Computer-designed commercial aircraft, June 12, 1994 - Computer engineered Boeing 777-200 first flown.

First Mostly Composite Commercial Airframe, 2009. The Boeing 787 Dreamliner is a long-range, mid-size wide-body, twin-engine jet airliner developed by Boeing Commercial Airplanes. It seats 210 to 290 passengers, depending on the variant. Boeing states that it is the company's most fuel-efficient airliner and the world's first major airliner to use composite materials for most of its construction. The 787 consumes 20% less fuel than the similarly-sized Boeing 767. Some of its distinguishing features include a four-panel windshield, noise-reducing chevrons on its engine nacelles, and a smoother nose contour. The airliner's maiden flight took place on December 15, 2009

First Scramjet Aircraft, 2005, A scramjet (*supersonic combustion ramjet*) is a variant of a ramjet airbreathing jet engine in which combustion takes place in supersonic airflow. As in ramjets, a scramjet relies on high vehicle speed to forcefully compress and decelerate the incoming air before combustion (hence *ram*jet), but whereas a ramjet decelerates the air to subsonic velocities before combustion, airflow in a scramjet is supersonic throughout the entire engine. This allows the scramjet to efficiently operate at extremely high speeds: theoretical projections place the top speed of a scramjet between Mach 12 and Mach 24, which is near orbital velocity. The fastest air-breathing aircraft is a SCRAM jet design, the NASA X-43A which reached Mach 9.8. For comparison, the second fastest air-breathing aircraft, the manned SR-71 Blackbird, has a cruising speed of Mach 3.2.

First successful flight test of Scramjet was performed by Russia in 1991. It was axisymmetric hydrogen-fueled dual-mode scramjet developed by the Central Institute of Aviation Motors, Moscow from late 1970s. The scramjet flight was flown captive-carry atop the SA-5 surface-to-air missile that included an experiment flight support unit known as the "Hypersonic Flying Laboratory" (HFL), "Kholod". Then during the period since 1992 up to 1998 an additional 6 flight tests of the axisymmetric high-speed scramjet-demonstrator were conducted by CIAM together with France and then with NASA, USA. Maximum flight velocity greater than Mach 6.4 was achieved and Scramjet operation during 77 seconds was demonstrated. These flight test series also provided insight into autonomous hypersonic flight controls.

Investigations of hypersonic flight have continued over the past few decades. At present, the U.S. military and NASA have formulated a "National Hypersonics Strategy" to investigate a range of options for hypersonic flight.

Different U.S. organizations have accepted hypersonic flight as a common goal. The U.S. Army desires hypersonic missiles that can attack mobile missile launchers quickly. NASA believes hypersonics could help develop economical, reusable launch vehicles. The Air Force is interested in a range of hypersonic systems, from air-launched cruise missiles to orbital spaceplanes, that the service believes could bring about a true "aerospace force."

The problem is complicated by the release of previously classified material and by partial publication, where claims are made, but specific parts of an experiment are kept secret. Additionally experimental difficulties in verifying that supersonic combustion actually occurred, or that actual net thrust was produced mean that at least four consortia have legitimate claims to "firsts", with several nations and institutions involved in each consortium. On June 15, 2007, the US Defense Advanced Research Agency (DARPA) and the Australian Defense Science and Technology Organization (DSTO), announced a successful scramjet flight at Mach 10 using rocket engines to boost the test vehicle to hypersonic speeds, at the Woomera Rocket Range in Central Australia.

Hypersonic X-43A

First Private craft to reach space, Oct 4, 2004 - Burt Rutan's *SpaceShipOne* in its second attempt from Mojave CA, entirely funded by private investment. The historic flight to 62 miles altitude by pilot Brian Binnie earned the $10 million Ansari-X prize.

First Global non-stop, non-refueled solo flight, Feb 28-Mar 3, 2005 - Steve Fossett in his GlobalFlyer [N277SF], produced by Burt Rutan team. Initial time estimate: 67h:02m eastward from and to Salina KS, covering nearly 23,000 statute miles. [1,3]

Life in the Universe and Where to find it

Astronautics

First Rockets, Ninth century Chinese Taoist alchemists discovered black powder while searching for the Elixir of life; this accidental discovery led to experiments in the form of weapons such as bombs, cannon, incendiary fire arrows and rocket-propelled fire arrows.

Exactly when the first flights of rockets occurred is contested. Some say that the first recorded use of a rocket in battle was by the Chinese in 1232 against the Mongol hordes. There were reports of fire arrows and 'iron pots' that could be heard for 5 leagues (25 km, or 15 miles) when they exploded upon impact, causing devastation for a radius of 600 meters (2,000 feet), apparently due to shrapnel. The lowering of the iron pots may have been a way for a besieged army to blow up invaders. The fire arrows were either arrows with explosives attached, or arrows propelled by gunpowder, such as the Korean Hwacha.

First Designs of Rocket for Space, In 1903, high school mathematics teacher Konstantin Tsiolkovsky (1857–1935) published *The Exploration of Cosmic Space by Means of Reaction Devices*, the first serious scientific work on space travel. The Tsiolkovsky rocket equation—the principle that governs rocket propulsion—is named in his honor (although it had been discovered previously). He also advocated the use of liquid hydrogen and oxygen as fuel, calculating their maximum exhaust velocity. His work was essentially unknown outside the Soviet Union, but inside the country it inspired further research, experimentation and the formation of the Society for Studies of Interplanetary Travel in 1924.

In 1912, Robert Esnault-Pelterie published a lecture on rocket theory and interplanetary travel. He independently derived Tsiolkovsky's rocket

278

equation, did basic calculations about the energy required to make round trips to the Moon and planets, and he proposed the use of atomic power (i.e. radium) to power a jet drive.

Robert Goddard began a serious analysis of rockets in 1912, concluding that conventional solid-fuel rockets needed to be improved in three ways. First, fuel should be burned in a small combustion chamber, instead of building the entire propellant container to withstand the high pressures. Second, rockets could be arranged in stages. And third, the exhaust speed (and thus the efficiency) could be greatly increased to beyond the speed of sound by using a De Laval nozzle. He patented these concepts in 1914. He also independently developed the mathematics of rocket flight. He proved that a rocket would work in a vacuum, which many scientists did not believe at the time.

In 1920, Goddard published these ideas and experimental results in *A Method of Reaching Extreme Altitudes*. The work included remarks about sending a solid-fuel rocket to the Moon, which attracted worldwide attention and was both praised and ridiculed. A New York Times editorial suggested that Professor Goddard: *"does not know of the relation of action to reaction, and the need to have something better than a vacuum against which to react—to say that would be absurd"* but that *"there are such things as intentional mistakes or oversights."*

Goddard's historical impact was diminished by the fact that he worked much in secret, though he offered his services to the military but was mostly ignored. This secrecy was prompted in part by his bad experience with the press and in part by his belief that his ideas were being plagiarized by foreign scientists. He was also in bad health and did not want to waste time helping amateurs and arguing with other scientists who did not understand this new science.

In 1923, Hermann Oberth (1894–1989) published "The Rocket into Planetary Space", a version of his doctoral thesis, after the University of Munich rejected it.

In 1924, Tsiolkovsky also wrote about multi-stage rockets, in 'Cosmic Rocket Trains'

First Long Range Rocket, V2 In 1943, production of the V-2 rocket began. The V-2 had an operational range of 300 km (185 miles) and carried a 1000 kg (2204 lb) warhead, with an amatol explosive charge. Highest point of altitude of its flight trajectory is 90 km. The vehicle was only different in details from most modern rockets, with turbo-pumps, inertial guidance and many other features. Thousands were fired at various Allied nations, mainly England, as well as Belgium and France. While they could not be intercepted, their guidance system design and single conventional warhead meant that the V-2 was insufficiently accurate against military targets. The later versions however, were more accurate, sometimes within meters, and could be devastating. 2,754 people in England were killed, and 6,523 were wounded before the launch campaign was terminated. While the V-2 did not significantly affect the course of the war, it provided a lethal demonstration of the potential for guided rockets as weapons.

Under *Projekt Amerika* Nazi Germany also tried to develop and use the first submarine launched ballistic missile (SLBMs) and the first intercontinental ballistic missiles (ICBMs) to bomb New York and other American cities. The tests of SLBM-variants of the A4 rocket was achieved with U-boat submarines towing launch platforms. The second stage of the A9/A10 rocket was tested a few times in January, February and March 1945.

In parallel with the guided missile program in Nazi Germany, rockets were also used on aircraft, either for assisting horizontal take-off (JATO), vertical take-off (Bachem Ba 349 "Natter") or for powering them (Me 163 etc.). During the war Germany also developed several guided and unguided air-to-air, ground-to-air and ground-to-ground missiles.

At the end of World War II, competing Russian, British, and U.S. military and scientific crews raced to capture technology and trained personnel from the German rocket program at Peenemunde. Russia and Britain had some success, but the United States benefited the most. The US captured a large number of German rocket scientists (many of whom were members of the Nazi Party, including von Braun) and brought them to the United States as part of Operation Paperclip. In America, the same rockets that were designed to rain down on Britain were used instead by scientists as research vehicles for developing the new technology further. The V-2 evolved into the American Redstone rocket, used in the early space program.

After the war, rockets were used to study high-altitude conditions, by radio telemetry of temperature and pressure of the atmosphere, detection of cosmic rays, and further research; notably for the Bell X-1 to break the sound barrier. This continued in the U.S. under Von Braun and the others, who were destined to become part of the U.S. scientific complex.

Independently, research continued in the Soviet Union under the leadership of the chief designer Sergei Korolev. With the help of German technicians, the V-2 was duplicated and improved as the R-1, R-2 and R-5 and missiles. German designs were abandoned in the late 1940s, and the foreign workers were sent home. A new series of engines built by Glushko and based on inventions of Aleksei Mihailovich Isaev formed the basis of the first ICBM, the R-7. The R-7 launched the first satellite- Sputnik, and later Yuri Gagarin -the first man into space, and the first lunar and planetary probes. This rocket is still in use today. These epoch marking events attracted the attention of top politicians, along with more money for further research.

First Satellite Sputnik 1 was the first artificial satellite to be put into Earth's orbit. It was launched into an elliptical low Earth orbit by the Soviet Union on 4 October 1957, and was the first in a series of satellites collectively known as the Sputnik program. The unanticipated announcement of *Sputnik 1*'s success precipitated the Sputnik crisis in the United States and ignited the Space Race within the Cold War . The launch ushered in new political, military, technological, and scientific developments. While the Sputnik launch was a single event, it marked the start of the Space Age. Apart from its value as a technological first, *Sputnik* also helped to identify the upper atmospheric layer's density, through measuring the satellite's orbital changes. It also provided data on radio-signal distribution in the ionosphere. Pressurized nitrogen in the satellite's body provided the first opportunity for meteoroid detection. If a meteoroid penetrated the satellite's outer hull, it would be detected by the temperature data sent back to Earth.

Sputnik 1 was launched during the International Geophysical Year from Site No.1, at the 5th Tyuratam range, in Kazakh SSR (now at the Baikonur Cosmodrome). The satellite travelled at 29,000 kilometers (18,000 mi) per hour, taking 96.2 minutes to complete an orbit, and emitted radio signals at 20.005 and 40.002 MHz which were monitored by amateur radio operators throughout the world. The signals continued for 22 days until the transmitter batteries ran out on 26 October 1957. *Sputnik 1* burned up on 4 January 1958, as it fell from orbit upon reentering Earth's atmosphere, after traveling about 60 million km (37 million miles) and spending 3 months in orbit. [1]

First Manned Space Flights, The first human spaceflight took place on April 12, 1961, when cosmonaut Yuri Gagarin made one orbit around the Earth aboard the Vostok 1 spacecraft, launched by the Societ space program and designed by the rocket scientist Sergey Korolyov. Valentina Tereshkova became the first woman in space on board Vostok 6 on June 16, 1963. Both spacecraft were launched by Vostok 3KA launch vehicles. Alexei Leonov made the first spacewalk when he left the Voskhod 2 on March 8, 1965. Svetlana Savitskaya became the first woman to do so on July 25, 1984.

The United State became the second nation to achieve manned spaceflight, with the suborbital flight of astronaut Alan Shepard aboard Freedom 7, carried out as part of Project Mercury. The spacecraft was launched on May 5, 1961 on a Redstone rocket. The first U.S. orbital flight was that of John Glenn aboard Friendship 7, which was launched February 20, 1962 on an Atlas rocket.

The People's Republic of China became the third nation to achieve human spaceflight when Yang Liwei launched into space on a Chinese-made vehicle, the Shenzhou 5, on October 15, 2003. The flight made China the third nation to have launched its own manned spacecraft using its own launcher. Previous European (Hermes) and Japanese (HOPE-X) domestic manned programs were abandoned after years of development, as was the first Chinese attempt, the Shuguang spacecraft.

Gemini NASA selected Mcdonnell Aircraft, which had been the prime contractor for the Project Mercury capsule, to build the Gemini capsule in 1961 and the first capsule was delivered in 1963. The spacecraft was 19 feet long and 10 feet wide with a launch weight of 8,490 pounds. The Gemini capsule first flew with a crew on March 23, 1965.
Gemini was the first manned spacecraft to include an onboard computer, the Gemini Guidance Computer, to facilitate management and control of mission maneuvers. Unlike the Mercury, it used ejection seats, in-flight radar and an artificial horizon—devices borrowed from the aviation industry.

A major difference between the Gemini and Mercury spacecraft was that Mercury had all systems other than the reentry rockets situated within the capsule, most of which were accessed through the astronaut's hatchway. In contrast, Gemini housed power, propulsion, and life support systems in a detachable Equipment Module located behind the Reentry Module. Many components in the capsule itself were reachable through their own small access doors.

The original intention was for Gemini to land on solid ground instead of at sea, using a Rogallo wing paraglider rather than a parachute, with the crew seated upright controlling the forward motion of the craft. To facilitate this, the paraglider did not attach just to the nose of the craft, but to an additional attachment point for balance near the heat shield. This cord was covered by a strip of metal which ran between the twin hatches. However, this design was ultimately dropped, and parachutes were used to make a sea landing as in Project Mercury. However, the capsule was suspended at an angle closer to horizontal, so that a side of the heat shield contacted the water first. This eliminated the need for the landing bag cushion used in the Mercury capsule.

Early short-duration missions had their electrical power supplied by batteries; later endurance missions used the first fuel cells in manned spacecraft.

Unlike Mercury, which could only change its orientation in space, the Gemini spacecraft could translate in all six directions, and alter its orbit. It was designed to dock with the Agena target vehicle, which had its own large rocket engine which was used to perform large orbital changes. [1]

Apollo The Command Module (CM) was the crew cabin, surrounded by a conical re-entry heat shield, designed to carry three astronauts from launch to lunar orbit and back to an Earth ocean splashdown. As such, it was the only component of the Apollo spacecraft to survive without major configuration changes as the program evolved from the early Apollo study designs. Equipment carried by the Command Module included reaction control engines, a docking tunnel, guidance and navigation systems and the Apollo Guidance Computer.

Attached to the Command Module was the cylindrical Service Module (SM), which housed the service propulsion engine and its propellants, the fuel cell power system, four maneuvering thruster quads, a high-gain S-band antenna for communications between the Moon and Earth, and storage tanks for water and oxygen. On the last three lunar missions, it also carried a scientific instrument package. Because its configuration was chosen early before the selection of lunar orbit rendezvous, the service propulsion engine was sized to lift the CSM off of the Moon, and thus oversized to about twice the thrust required for translunar flight.

As used in the actual lunar program, the two modules remained attached throughout most of the flight to make a single ferry craft, somewhat awkwardly known as the Command/Service Module (CSM) which carried a separate lunar lander (only half as heavy as the CSM) to the Moon, and the astronauts home to Earth. Just before re-entry, the Service Module was discarded and only the Command Module re-entered the atmosphere, using its heat shield to survive the intense heat caused by air friction. After re-entry it deployed parachutes that slowed its descent, allowing a smooth splashdown in the ocean.

Under the leadership of Harrison Storms, North American Aviation won the contract to build the CSM, and also the second stage of the Saturn V launch vehicle for NASA. Relations between North American and NASA were strained during the winter of 1965-66 by delivery delays, quality shortfalls, and cost overruns in both components. They were strained even more a year later when a cabin fire killed the crew of Apollo 1 during a ground test. The cause was determined to be an electrical short in the wiring of the Command Module;

while the determination of responsibility for the accident was complex, the review board concluded that "deficiencies existed in Command Module design, workmanship and quality control." This eventually led to the removal of Storms as Command Module program manager.

The Lunar Module (LM), originally known as the Lunar Excursion Module, or LEM, was designed to fly between lunar orbit and the surface, landing two astronauts on the Moon and taking them back to the Command Module. It had no aerodynamic heat shield and was of a construction so lightweight that it would not have been able to fly through the Earth's atmosphere. It consisted of two stages, a descent and an ascent stage. The descent stage contained compartments which carried cargo such as the Apollo Lunar Surface Experiment Package and Lunar Rover.

The contract for design and construction of the Lunar Module was awarded to Grumman Aircraft Engineering Corporation, and the project was overseen by Tom Kelly. There were also problems with the Lunar Module; due to delays in the test program, the LM became a "pacing item," meaning that it was in danger of delaying the schedule of the whole Apollo program. The first manned LM was not ready for its planned Earth orbit test in December 1968, but the program was kept on schedule by canceling a second manned Earth orbit LM flight. [1]

Shuttle The Space Shuttle was the first operational orbital spacecraft designed for reuse. It carried different payloads to low Earth orbit, provided crew rotation for the International Space Station (ISS), and performed servicing missions. The orbiter could also recover satellites and other payloads from orbit and return them to Earth. Each Shuttle was designed for a projected lifespan of 100 launches or ten years of operational life, although this was later extended. The person in charge of designing the STS was Maxime Faget, who had also overseen the Mercury, Gemini, and Apollo spacecraft designs. The crucial factor in the size and shape of the Shuttle Orbiter was the requirement that it be able to accommodate the largest planned commercial and military satellites, and have the cross-range recovery range to meet the requirement for classified USAF missions for a once-around abort from a launch to a polar orbit. Factors involved in opting for solid rockets and an expendable fuel tank included the desire of the Pentagon to obtain a high-capacity payload vehicle for satellite deployment, and the desire of the Nixon administration to reduce the costs of space exploration by developing a spacecraft with reusable components.

Each Space Shuttle is a reusable launch system that is composed of three main assemblies: the reusable Orbiter Vehicle (OV), the expendable external tank (ET), and the two reusable solid rocket boosters (SRBs). Only the orbiter entered orbit shortly after the tank and boosters are jettisoned. The vehicle was launched vertically like a conventional rocket, and the orbiter glided to a horizontal landing like an airplane, after which it was refurbished for reuse. The SRBs parachuted to splashdown in the ocean where they were towed back to shore and refurbished for later shuttle missions.

Five space-worthy orbiters were built: Columbia (OV-102), Challenger (OV-099), Discovery (OV-103), Atlantis (OV-104), and Endeavour (OV-105). An additional craft, Enterprise (OV-101), was not built for orbital space flight, and was used only for testing gliding and landing. *Enterprise* was originally intended to be made fully space-worthy after use for the approach and landing test (ALT) program, but it was found more economical to upgrade the structural test article STA-099 into orbiter *Challenger* (OV-099). *Challenger* disintegrated 73 seconds after launch in 1986, and *Endeavour* was built as a replacement for Challenger from structural spare components. *Columbia* broke apart during re-entry in 2003. Building Space Shuttle Endeavour cost about US$1.7 billion. One Space Shuttle launch costs around $450 million.

Roger A. Pielke, Jr. has estimated that the Space Shuttle program has cost about US$170 billion (2008 dollars) through early 2008. This works out to an average cost per flight of about US$1.5 billion. However, two missions were paid for by Germany, Spacelab D1 and D2 (D for *Deutschland*) with a payload control center in Oberpfaffenhofen, Germany. D1 was the first time that control of a manned STS mission payload was not in U.S. hands.

ISS The International Space Station (ISS) is an internationally-developed research facility assembled in low Earth orbit and is the largest space station ever constructed. On-orbit construction of the station began in 1998 and is expected to be finished in 2012. The station is expected to remain in operation until at least 2020, and potentially to 2028. Like many artificial satellites, the ISS can be seen from Earth with the naked eye. The ISS serves as a research laboratory that has a microgravity environment in which crews conduct experiments in biology, human biology, physics, astronomy and meteorology. The station has a unique environment for the testing of the spacecraft systems that will be required for missions to the Moon and Mars. The ISS is operated by Expedition crews, and has been continuously staffed since 2 November 2000—an uninterrupted human presence in space for the

289

past 10 years and 308 days. As of June 2011, the crew of Expedition 28 is aboard.

The ISS is a synthesis of several space station projects that includes the American Freedom, the Soviet/Russian Mir-2, the European *Columbus* and the Japanese *Kibo*. Budget constraints led to the merger of these projects into a single multi-national program. The ISS project began in 1994 with the Shuttle-Mir program and the first module of the station, Zarya, was launched in 1998 by Russia. Since then, pressurized modules, external trusses and other components have been launched by American space shuttles, Russian Proton rockets and Russian Soyuz rockets. As of June 2011, the station consisted of 15 pressurized modules and an extensive integrated truss structure (ITS). The planned final module, the Russian laboratory module, is expected to launch in 2012. Power is provided by 16 solar arrays mounted on the external truss, in addition to four smaller arrays on the Russian modules. The station is maintained at an orbit between 278 km (173 mi) and 460 km (286 mi) altitude, and travels at an average ground speed of 27,724 km (17,227 mi) per hour, completing 15.7 orbits per day.

Operated as a joint project between the five participant space agencies, the station's sections are controlled by mission control centers on the ground operated by the American National Aeronautics and Space Administration (NASA), the Russian Federal Space Agency (RKA), the Japan Aerospace Exploration Agency (JAXA), the Canadian Space Agency (CSA), and the European Space Agency (ESA). The ownership and use of the space station is established in intergovernmental treaties and agreements that allow the Russian Federation to retain full ownership of its own modules, with the remainder of the station allocated between the other international partners. The station is serviced by Soyuz spacecraft, Progress spacecraft, the

Automated Transfer Vehicle and the H-II Transfer Vehicle and has been visited by astronauts and cosmonauts from 15 different nations. The cost of the station has been estimated by ESA as €100 billion over 30 years, although other estimates range from 35 billion dollars to 160 billion dollars. The financing, research capabilities and technical design of the ISS program have been criticized because of the high cost. [1]

Astronomy

2500 BC   Many ancient sites are thought to have astronomical significance, such as the Ancient Egyptian pyramids, Harappan shell instruments, British megaliths, and buildings in China and.

750 BC   Babylonian astronomers discover 18.6-year cycle in the rising and setting of the Moon. From this they created the first almanacs - tables of the movements of the Sun, Moon and planets for the use in astrology. In 6th century Greece, this knowledge is used to predict eclipses.

585 BC   Thales predicts a solar eclipse.

388 BC   Plato, a Greek philosopher, founds a school (the Platonic Academy) that will influence the next 2000 years. This promotes the idea that everything in the universe moves in harmony and that the Sun, Moon, and planets move around Earth in perfect circles.

270 BC   Aristarchus of Samos proposes heliocentrism as an alternative

to the Earth-centered universe. His heliocentric model places the Sun at its center, with Earth as just one planet orbiting it. However, there were only a few people who took the theory seriously.

164 BC — The earliest recorded sighting of Halley's comet is made by Babylonian astronomers. Their records of the comet's movement allow astronomers today to predict accurately how the comet's orbit changes over the centuries.

150 AD — Ptolemy publishes his star catalogue, listing 48 constellations and endorses the geocentric (Earth-centered) view of the universe. His views go unquestioned for nearly 1500 years in Europe, and are passed down to Arabic and medieval European astronomers in his book the *Almagest*.

400 — The Hindu cosmological time cycles explained in the *Surya Siddhanta*, gives the average length of the sidereal year (the length of the Earth's revolution around the Sun) as 365.2563627 days, which is only 1.4 seconds longer than the modern value of 365.2563627 days. This remains the most accurate estimate for the length of the sidereal year anywhere in the world for over a thousand years.

499 — Indian mathematician-astronomer Aryabhata, in his *Aryabhatiya*, propounds a heliocentric solar system of gravitation, and an eccentric elliptical model of the planets, where the planets spin on their axes and follow elliptical orbits around the Sun. He also writes that the planets and the Moon do not have their own light but reflect the light of the Sun, and

that the Earth rotates on its axis causing day and night and also round the sun causing year. Aryabhata gives the radii of planetary orbits in terms of orbit of earth/sun. Incredibly, he also believes that the orbits of the planets are ellipses and not circles, and also correctly explains the causes of eclipse of sun and moon. His calculation of Earth's diameter at 13383 km (8316 miles) would remain the most accurate approximation for over a thousand years. Aryabhata also accurately computes the Earth's circumference, the solar and lunar eclipses, and the length of Earth's revolution around the Sun.

628 Indian mathematician-astronomer Brahmagupta, in his *Brahma-Sphuta-Siddhanta*, first recognizes gravity as a force of attraction, and briefly describes the law of gravitation. He gives methods for calculations of the motions and places of various planets, their rising and setting, conjunctions, and calculations of the solar and lunar eclipses.

773 The Sanskrit works of Aryabhata and Brahmagupta, along with the Sanskrit text *Surya Siddhanta*, are translated into Arabic, introducing Arabic astronomers to Indian astronomy.

777 Muhammad al-Fazari and Yaqub ibn Tariq translate the *Surya Siddhanta* and *Brahmasphutasiddhanta*, and compile them as the *Zij al-Sindhind*, the first Zij treatise.

830 The first major Arabic work of astronomy is the *Zij al-Sindh* by al-Khwarizimi. The work contains tables for the movements of the sun, the moon and the five planets known at the time.

The work is significant as it introduced Ptolemaic concepts into Islamic sciences. This work also marks the turning point in Arabic astronomy. Hitherto, Arabic astronomers had adopted a primarily research approach to the field, translating works of others and learning already discovered knowledge. Al-Khwarizmi's work marked the beginning of nontraditional methods of study and calculations.

850   al-Farghani wrote *Kitab fi Jawani* ("*A compendium of the science of stars*"). The book primarily gave a summary of Ptolemic cosmography. However, it also corrected Ptolemy based on findings of earlier Arab astronomers. Al-Farghani gave revised values for the obliquity of the ecliptic, the precessional movement of the apogees of the sun and the moon, and the circumference of the earth. The books were widely circulated through the Muslim world, and even translated into Latin.

928   The earliest surviving astrolabe is constructed by Islamic mathematician- astronomer Mohammad al-Fazari. Astrolabes are the most advanced instruments of their time. The precise measurement of the positions of stars and planets allows Islamic astronomers to compile the most detailed almanacs and star atlases yet.

1030   Abu al-Rayhan al-Biruni discussed the Indian heliocentric theories of Aryabhata, Brahmagupta and Varahamihira in his *Ta'rikh al-Hind* (*Indica* in Latin). Biruni stated that the followers of Aryabhata consider the Sun to be at the center. In fact,

Biruni casually stated that this does not create any mathematical problems.

1031   Abu Said Sinjari, a contemporary of Abu Rayhan Biruni, suggested the possible heliocentric movement of the Earth around the Sun.

1054   Chinese astronomers record the sudden appearance of a bright star. Native-American rock carvings also show the brilliant star close to the Moon. This star is the Crab supernova exploding.

1070   Abu Ubayd al-Juzjani published the *Tarik al-Aflak*. In his work, he indicated the so-called "equant" problem of the Ptolemic model. Al-Juzjani even proposed a solution for the problem. In al-Andalus, the anonymous work *al-Istidrak ala Batlamyus* (meaning "Recapitulation regarding Ptolemy"), included a list of objections to the Ptolemic astronomy.

One of the most important works in the period was *Al-Shuku ala Batlamyus* ("*Doubts on Ptolemy*"). In this, the author summed up the inconsistencies of the Ptolemic models. Many astronomers took up the challenge posed in this work, namely to develop alternate models that evaded such errors.

1126   Islamic and Indian astronomical works (including *Aryabhatiya* and *Brahma-Sphuta-Siddhanta*) are translated into Latin in Cordoba, Spain in 1126, introducing European astronomers to Islamic and Indian astronomy.

1150   Indian mathematician-astronomer Bhaskara II, in his

Siddhanta Shiromani, calculates the longitudes and latitudes of the planets, lunar and solar eclipses, risings and settings, the Moon's lunar crescent, syzygies, and conjunctions of the planets with each other and with the fixed stars, and explains the three problems of diurnal rotation. He also calculates the planetary mean motion, ellipses, first visibilities of the planets, the lunar crescent, the seasons, and the length of the Earth's revolution around the Sun to 9 decimal places.

1250   Mo'ayyeduddin Urdi develops the Urdi lemma, which is later used in the Copernican heliocentric model. Nasir al-Din al-Tusi resolved significant problems in the Ptolemaic system by developing the Tusi-couple as an alternative to the physically problematic equant introduced by Ptolemy. His Tusi-couple is later used in the Copernican model. Tusi's student Qutb al-Din al-Shirazi, in his *The Limit of Accomplishment concerning Knowledge of the Heavens*, discusses the possibility of heliocentrism. Najm al-Din al-Qazwini al-Katibi, who also worked at the Maraghah observatory, in his *Hikmat al-'Ain*, wrote an argument for a heliocentric model, though he later abandoned the idea.

1350   Ibn al-Shatir (1304–1375), in his *A Final Inquiry Concerning the Rectification of Planetary Theory*, eliminated the need for an equant by introducing an extra epicycle, departing from the Ptolemaic system in a way very similar to what Copernicus later also did. Ibn al-Shatir proposed a system that was only approximately geocentric, rather than exactly so, having demonstrated trigonometrically that the Earth was not the exact center of the universe. His rectification is later used in the Copernican model.

1543   Nicolaus Copernicus publishes *De revolutionbus orbium coelestium* containing his theory that Earth travels around the Sun. However, he complicates his theory by retaining Plato's perfect circular orbits of the planets.

1572   A brilliant supernova (SN 1572 - thought, at the time, to be a comet) is observed by Tycho Brahe, who proves that it is traveling beyond Earth's atmosphere and therefore provides the first evidence that the heavens can change.

1608   Dutch eyeglass maker Hans Lippershey invents the refracting telescope. The invention spreads rapidly across Europe, as scientists make their own instruments. Their discoveries begin a revolution in astronomy.

1609   Johannes Kepler publishes his *New Astronomy*. In this and later works, he announces his three laws of planetary motion, replacing the circular orbits of Plato with elliptical ones. Almanacs based on his laws prove to be highly accurate.

1610   Galileo Galilei publishes *Sidereus Nuncius* describing the findings of his observations with the telescope he built. These include spots on the Sun. craters on the Moon, and four satellites of Jupiter. Proving that not everything orbits Earth, he promotes the Copernican view of a Sun-centered universe.

1655   As the power and the quality of the telescopes increases, Christiaan Huygens studies Saturn and discovers its largest satellite, Titan. He also explains Saturn's appearance, suggesting the planet is surround by a thin ring.

1663   Scottish astronomer James Gregory describes his "gregorian" reflecting telescope, using parabolic mirrors instead of lenses to reduce chromatic aberration and spherical aberration, but is unable to build one.

1668   Isaac Newton 1668 builds the first reflecting telescope, his Newtonian telescope.

1687   Isaac Newton publishes his *Philosophiae Naturalis Principia Mathematica*, establishing the theory of gravitation and laws of motion. The *Principia* explains Kepler's laws of planetary motion and allows astronomers to understand the forces acting between the Sun, the planets, and their moons.

1705   Edmond Halley calculates that the comets recorded at 76-year intervals from 1456 to 1682 are one and the same. He predicts that the comet will return again in 1758. When it reappears as expected, the comet is named in his honor.

1750   French astronomer Nicolas de Lacaille sails to southern oceans and begins work compiling a catalog of more than 10000 stars in the southern sky. Although Halley and others have observed from the Southern Hemisphere before, Lacaille's star catalog is the first comprehensive one of the southern sky.

1781   Amateur astronomer William Hershel discovers the planet Uranus, although he at first mistakes it for a comet. Uranus is the first planet to be discovered beyond Saturn, which was thought to be the most distant planet in ancient times.

1784   Charles Messier publishes his catalog of star clusters and nebulas. Messier draws up the list to prevent these objects from being identified as comets. However, it soon becomes a standard reference for the study of star clusters and nebulas and is still in use today.

1800   William Herschel splits sunlight through a prism and with a thermometer, measures the energy given out by different colors. He notices a sudden increase in energy beyond the red end of the spectrum, discovering invisible infrared and laying the foundations of spectroscopy.

1801   Italian astronomer Giuseppe Piazzi discovers what appears to be a new planet orbiting between Mars and Jupiter, and names it Ceres. William Herschel proves it is a very small object, calculating it to be only 320 km in diameter, and not a planet. He proposes the name asteroid, and soon other similar bodies are being found. We now know that Ceres is 932 km in diameter, however, it is still too small to be a planet.

1814   Joseph von Fraunhofer builds the first accurate spectrometer and uses it to study the spectrum of the Sun's light. He discovers and maps hundreds of fine dark lines crossing the solar spectrum. In 1859 these lines are linked to chemical elements in the Sun's atmosphere. Spectroscopy becomes a method for studying what stars are made of.

Life in the Universe and Where to find it

1838   Friedrich Bessel successfully uses the method of stellar parallax, the effect of Earth's annual movement around the Sun, to calculate the distance to 61 Cygni: the first star other that the Sun to have its distance measured. Bessel has pioneered the truly accurate measurement of stellar positions, and the parallax technique establishes a framework for measuring the scale of the universe.

1843   German Amateur astronomer Heinrich Schwabe, who has been studying the Sun for the past 17 years, announces his discovery of a regular cycle in sunspot numbers - the first clue to the Sun's internal structure.

1845   Irish astronomer William Parsons, 3rd Earl of Rosse completes the first of the world's great telescopes, with a 180-cm mirror. He uses it to study and draw the structure of nebulas, and within a few months discovers the spiral structure of the Whirlpool Galaxy.

French physicists Jean Foucault and Armand Fizeau take the first detailed photographs of the Sun's surface through a telescope - the birth of scientific astrophotography. Within five years, astronomers produce the first detailed photographs of the Moon. Early film is not sensitive enough to image stars.

1846   A new planet, Neptune, is identified by German astronomer Johann Gottfried Galle while searching in the position suggested by Urbain Le Verrier. Le Verrier has calculated the position and size of the planet from the effects of its gravitational pull on the orbit of Uranus. An English mathematician, John Couch Adams, also made a similar calculation a year earlier.

1868   Astronomers notice a new bright emission line in the spectrum of the Sun's atmosphere during an eclipse. The emission line is caused by an element's giving out light, and British astronomer Norman Lockyer concludes that it is an element unknown on Earth. He calls it helium, from the Greek word for the Sun. Nearly 30 years later, helium is found on Earth.

1872   An American astronomer Henry Draper takes the first photograph of the spectrum of a star (Vega), showing absorption lines that reveal its chemical makeup. Astronomers begin to see that spectroscopy is the key to understanding how stars evolve. William Huggins uses absorption lines to measure the redshifts of stars, which give the first indication of how fast stars are moving.

1895   Konstantin Tsiolkovsky publishes his first article on the possibility of space flight. His greatest discovery is that a rocket, unlike other forms of propulsion, will work in a vacuum. He also outlines the principle of a multistage launch vehicle.

1901   A comprehensive survey of stars, the Henry Draper Catalog, is published. In the catalog, Anne Jump Cannon proposes a sequence of classifying stars by the absorption lines in their spectra, which is still in use today.

1906   Ejnar Hertzsprung establishes the standard for measuring the true brightness of a star. He shows that there is a relationship between color and absolute magnitude for 90% of the stars in the Milky Way Galaxy. In 1913, Henry Norris Russell publishes a diagram that shows this relationship. Although astronomers agree that the diagram shows the sequence in which stars evolve, they argue about which way the sequence progresses. Arthur Eddington finally settles the controversy in 1924,

1916   German physicist Karl Schwarzschild uses Albert Einstein's theory of general relativity to lay the groundwork for black hole theory. He suggests that if any star collapse to a certain size or smaller, its gravity will be so strong that no form of radiation will escape from it.

1923   Edwin Hubble discovers a Cepheid variable star in the "Andromeda Nebula" and proves that Andromeda and other nebulas are galaxies far beyond our own. By 1925, he produces a classification system for galaxies.

1926   Robert Goddard launches the first rocket powered by liquid fuel. He also demonstrates that a rocket can work in a vacuum. His later rockets break the sound barrier for the first time.

1929   Edwin Hubble discovers that the universe is expanding and that the farther away a galaxy is, the faster it is moving away from us. Two years later, Georges Lemaitre suggests that the expansion can be traced to an initial "Big Bang".

1930   By applying new ideas from subatomic physics, Subrahmanyan Chandrasekhar predicts that the atoms in a white dwarf star of more than 1.44 solar masses will disintegrate, causing the star to collapse violently. In 1933, Walter Baade and Fritz Zwicky describe the neutron star that results from this collapse, causing a supernova explosion.

Clyde Tombaugh discovers the planet Pluto at the Lowell Observatory in Flagstaff, Arizona. The planet is so faint and moving so slowly that he has to compare photos taken several nights apart.

1932   Karl Jansky detects the first radio waves coming from space. In 1942, radio waves from the Sun are detected. Seven years later radio astronomers identify the first distant source - the Crab Nebula, and the galaxies Centaurus A and M87.

1938   German physicist Hans Bethe explains how stars generate energy. He outlines a series of nuclear fusion reactions that turn hydrogen into helium and release enormous amounts of energy in a star's core. These reactions use the star's hydrogen very slowly, allowing it to burn for billions of years.

1944   A team of German scientists led by Wernher von Braun develops the V-2, the first rocket-powered ballistic missile. Scientists and engineers from Braun's team were captured at the end of World War II and drafted into the American and Russian rocket programs.

1948   The largest telescope in the world, with a 5.08m (200 in) mirror, is completed at Palomar Mountain in California. At the time, the telescope pushes single-mirror telescope technology to its limits - large mirrors tend to bend under their own weight.

1957   Russia launches the first artificial satellite, Sputnik 1, into orbit, beginning the space age. The US launches its first satellite, Explorer 1, four months later.

1959   Russia and the US both launch probes to the Moon, but NASA's Pioneer probes all failed. The Russian Luna program was more successful. Luna 2 crash-lands on the Moon's surface in September, and Luna 3 returns the first pictures of the Moon's farside in October.

1961   Russia takes the lead in the space race as Yuri Gagarin becomes the first person to orbit Earth in April. NASA astronaut Alan Shepard becomes the first American in space a month later, but does not go into orbit. John Glenn achieves this in early 1962.

1962   Mariner 2 becomes the first probe to reach another planet, flying past Venus in December. NASA follows this with the successful Mariner 4 mission to Mars in 1965, both the US and Russia sends many more probes to planets through the rest of the 1960s and 1970s.

1963   Dutch-American astronomer Maarten Schmidt measures the spectra of quasars, the mysterious starlike radio sources discovered in 1960. He establishes that quasars are active galaxies, and among the most distant objects in the universe.

1965   Arno Penzias and Robert Wilson announce the discovery of a weak radio signal coming from all parts of the sky. Scientists figure out that this must be emitted by an object at a temperature of -270°C. Soon it is recognized as the remnant of the very hot radiation from the Big Bang that created the universe 13 billion years ago.

1966   Russian Luna 9 probe makes the first successful soft landing on the Moon in January, while the US lands the far more complex Surveyor missions, which follows up to NASA's Ranger series of crash landers, scout sites for possible manned landings.

1967   Jocelyn Bell Burnell and Antony Hewish detected the first pulsar, an object emitting regular pulses of radio waves. Pulsars are eventually recognized as rapidly spinning neutron stars with intense magnetic fields - the remains of a supernova explosion.

1969   The US wins the race for the Moon, as Neil Armstrong steps onto the lunar surface on July 20. Apollo 11 is followed by five further landing missions, three carrying a sophisticated lunar rover vehicle.

1970   The Uhuru satellite, designed to map the sky at X-ray wavelengths, is launched by NASA. The existence of X rays from the Sun and a few other stars has already been found using rocket-launched experiments, but Uhuru charts more than 300 X-ray sources, including several possible black holes.

Life in the Universe and Where to find it

1970   C.T. Bolton was the first to develop a computer model for stellar atmospheres.

1971   Russia launches its first space station, Salyut 1, into orbit. It is followed by a series of stations, culminating with Mir in 1986. A permanent platform in orbit allows cosmonauts to carry out serious research and to set a series of new duration records for spaceflight.

1972   Charles Thomas Bolton was the first astronomer to present irrefutable evidence of the existence of a black hole.

1975   The Russian probe Venera 9 lands on the surface of Venus and sends back the first picture of its surface. The first probe to land on another planet, Venera 7 in 1970, had no camera. Both break down within an hour in the hostile atmosphere.

1976   Two NASA probes arrive at Mars. Each Viking mission consists of an orbiter, which photographs the planet from above, and a lander, which touches down on the surface, analyzes the rocks, and searches unsuccessfully for life.

1977   Two Voyager probes are launched by NASA to the outer planets. The Voyagers return scientific data and pictures from Jupiter and Saturn, and, before leaving the solar system, Voyager 2 becomes the first probe to visit Uranus an Neptune.

1981   Columbia, the first of NASA's reusable space shuttles, makes its maiden flight, ten years in development, the shuttle will make space travel routine and eventually open the path for a new International Space Station.

306

1983   The first infrared astronomy satellite, ITAS, is launched. It must be cooled to extremely low temperatures with liquid helium, and it operates for only 300 days before the supply of helium is exhausted. During this time it completes an infrared survey of 98% of the sky.

1986   NASA's spaceflight program comes to a halt when the space shuttle Challenger explodes shortly after launch. A thorough inquiry and modifications to the rest of the fleet kept the shuttles on the ground for nearly three years.

        The returning Halley's comet is met by a fleet of five probes from Russia, Japan, and Europe. The most ambitious is the European Space Agency's Giotto, which flies through the comet's coma and photographs the nucleus.

1990   The Magellan probe, launched by NASA, arrives at Venus and spends three years mapping the planet with radar. Magellan is the first in a new wave of probes that include Galileo, which arrives at Jupiter in 1995, and Cassini which arrives at Saturn in 2004.

        The Hubble Space Telescope, the first large optical telescope in orbit, is launched using the space shuttle, but astronomers soon discovered that it is crippled by a problem with its mirror. A complex repair mission in 1993 allows the telescope to start producing spectacular images of distant stars, nebulas, and galaxies.

1992   The Cosmic Background Explorer satellite produces a detailed map of the background radiation remaining from the Big Bang.

The map shows "ripples", caused by slight variations in the density of the early universe - the seeds of galaxies and galaxy clusters.

The 10-m Keck telescope on Mauna Kea, Hawaii, is completed. The first revolutionary new wave of telescopes, the Keck's main mirror is made of 36 six-sided segments, with computers to control their alignment. New optical telescopes also make use of interferometry - improving resolution by combining images from separate telescopes.

1998   Construction work on a huge new space station named ISS is begun. A joint venture between many countries, including former space rivals Russia and the US. [1]

Computing

2400 BC   Abacus: The abacus, the first known calculator, was invented in Babylonia.

500 BC   Panini: Introduced the forerunner to modern formal language theory

300 BC   Pingala: Pingala invented the binary number system

87 BC   Antikythera Mechanism: Built in Rhodes to track movement of the stars

60 BC   Heron of Alexandria: Heron of Alexandria invents machines which follow a series of instructions

724   Liang Ling-Can invents the first fully mechanical clock

1492   Drawings by Leonardo da Vinci depict inventions such as flying machines, including a helicopter, the first mechanical calculator and one of the first programmable robots

1614   John Napier invents a system of moveable rods (Napier's Rods) based on logarithms which was able to multiply, divide and calculate square and cube roots.

1622   Willim Oughtred develops slide rules

1623   Wilhelm Schickard invents calculating clock

1642   Blaise Pascal invents the "Pascaline", a mechanical adding machine

1671   Gottfried Leibniz is known as one of the founding fathers of calculus, along with Isaac Newton

1801   Joseph-Marie Jacquar invents an automatic loom controlled by punched cards

1820   The Arithmometer was first mass-produced calculator invented by Charles Xavier Thomas de Colmar

1822	Charles Babbage designs his first mechanical computer
1834	Charles Babbage invents the Analytical Engine
1835	Samuel Morse invents Morse code
1848	George Boole invents Boolean algebra
1858	Per Georg Scheutz and his son Edvard invent the Tabulating Machine
1869	William Stanley Jevons designs a practical logic machine
1878	Ramon Verea invents a fast calculator with an internal multiplication table
1880	Alexander Graham Bell invents the telephone called the Photophone
1884	Dorr E. Felt invents the Comptometer which is operated by pressing keys
1890	Herman Hollerith invents a counting machine which has an incremental mechanical counter
1895	Guglielmo Marconi "invents" radio signals

1896	Herman Hollerith forms the Tabulating Machine Company which later becomes IBM
1898	Nikola Tesla invents remote control
1906	Lee De Forest invents the electronic tube
1911	IBM formed on June 15, 1911
1923	Philo Farnsworth invents Television
1924	John Logie Baird invents Electro Mechanical television
1930	Vannevar Bush develops a partly electronic Difference Engine
1931	Kurt Godel publishes a paper on the use of a universal formal language
1937	Alan Turing develops the concept of a theoretical computing machine
1938	Konrad Zuse creates the Z1 Computer a binary digital computer using punch tape
1939	George Stibitz develops the Complex Number Calculator – a foundation for digital computers William Hewlett and David Packard start Hewlett Packard John Vincent Atanasoff and Clifford Berry develop the ABC

Life in the Universe and Where to find it

(Atanasoft-Berry Computer) prototype

1943   Adolf Hitler uses the Enigma encryption machine
         Alan turing develops the code breaking machine Colossus

1944   Howard Aiken and Grace Hopper design the MARK series of
         computers at Harvard University

1945   John Presper Eckert and John W. Mauchly develop the ENIAC
         (Electronic Numerical Integrator and Computer
         The term computer "bug" was first used by Grace Hopper who
         is also credited with the phrase "Its easier to ask for
         forgiveness than to get permission"

1946   F.C. Williams develops his cathode ray tube (CRT) storing
         deice which was the forerunner to random access memory
         (RAM)

1947   Donald Watts Davies joins Alan turing to build the fastest
         digital computer in England at the time, the Pilot ACE
         William Shockley invents the transistor at Bell Labs
         Douglas Engelbart theorizes on interactive computing with
         keyboard and screen display instead of on punchcards

1948   Andrew donald Booth invents magnetic drum memory
         Frederic Calland Williams and Tom Kilburn develop the SSEM
         "Small Scale Experimental Machine" digital CRT storage which
         was soon nicknamed the "Baby"

1949   Claude Shannon builds the first machine that plays chess
            Howard Aiken develops the Harvard-MARK III

1950   Hideo Yamchito creates the first electronic computer in Japan
            Alan Turing publishes his paper Computing Machinery and
            intelligence which helps create the Turing Test

1951   T. Ramond Thompson and John Simmons develop the first
            business computer, the Lyons Electronic Office (LEO) at Lyons
            Co.
            UNIVAC I (UNIVersal Automatic Computer I) was introduced,
            the first commercial computer made in the United States and
            designed principally by John Presper Eckert an John W.
            Mauchly
            EDVAC (Electronic Discrete Variable Automatic Computer)
            begins performin basic tasks. Unlike the ENIAC, it was binary
            rather than decimal

1953   The IBM 701 becomes available and a total of 19 are sold to
            the scientific community

1954   John Backus and IBM develop the FORTRAN Computer
            Programming Language

1955   Bell Labs introduces its first transistor computer

1956   Optical fiber was invented by Basil Hirschowitz, C. Wilbur

Life in the Universe and Where to find it

Peters and Lawrence E. Curtiss

1957   Sputnik I and Sputnik II are launched by the Russians

1958   ARPA (Advanced Research Projects Agency) and NASA are
            formed
            The first integrated circuit, or silicon ship, is produced by the
            US Jack Kilby and Robert Noyce

1959   Paul Baran theorizes on the "survivability of communication
            systems under nuclear attack", digital technology and
            symbiosis between humans and machines

1960   The common Business Oriented Language (COBOL)
            programming language is invented

1961   General Motors puts the first industrial robot, Unimate, to work
            in a New Jersey factory

1962   The first computer game Spacewar invented by Steve Russell
            and MIT

1963   Douglas Engelbart invents and patents the first computer
            mouse
            ASCII is developed to standardize data exchange among
            computers

1964   IBM introduces the first word processor

John Kemeny and Thomas Kurtz develop BASIC

1965   Andries van Dam and Ted Nelson coin the term "hypertext"

1967   IBM creates the first floppy disk

1969   Seymour Cray develops the CDC 7600, the first
        supercomputer
        Gary Starkweather invents the laser printer while working at
        Xerox
        The U.S. Department of Defense sets up the Advanced
        research Projects Agency Network (ARPANET) this network
        was the first building block to what the internet is today but
        originally with the intention of creating a computer network
        that could withstand any type of disaster.

1970   Intel introduces the world's first available dynamic RAM
        (Random Access Memory) chip and the first microprocessor,
        the Intel 4004

1971   E-mail was invented by Ray Tomlinson
        Liquid Crystal Display (LCD) invented by James Fergason
        Pocket calculator invented by Sharp Corporation

1972   Atari releases Pong, the first commercial video game
        The compact disc (CD) is invented in the U.S.

1973   Robert Metcalfe creates the Ethernet, a local area network

(LAN)

The minicomputer Xerox Alto was a landmark development in the devlopment of personal computers

Vint Cerf and Bob Kahn develop gateway routing computers to negotiate between the various national networks

1974   IBM develops SEQUEL (Structured English Query Language)

1975   Altair produces the first portable computer
The Microsoft Corporation was founded by Bill Gates and Paul Allen to develop and sell BASIC interpreters for the Altair 8800

1976   Apple Computers was founded by Steve Wozniak and Steve Jobs

1977   Apple Computer's Apple II, the first computer with color graphics is demonstrated
MODEM program is written allowing computers to communicate over telephone lines

1978   The first magnetic tape is developed in the U.S.

1979   Over half a million computers are in use in the U.S.

1980   IBM hires Paul Allen and Bill Gates to create an operating system for a new PC, which becomes DOS

1982	WordPerfect is introduced
	Commodore 64 becomes the best selling computer
1983	More than 10 million computers are in use in the U.S.
	DNS is introduced, domain names are introduced
1984	Apple introduces the Macintosh
	The term cyberspace is introduced by William Gibson in *Neuromancer*
1985	Pagemaker is introduced
	Nintendo Entertainment System makes its debut
1986	More than 30 million computers are in use in the U.S.
1987	Microsoft Works is introduced
	Perl is introduced
1988	Over 45 million PCs are in use in the U.S.
1990	The Internet, World Wide Web is designed to use hypertext
1991	The World Wide Web is introduced to the public
1993	50 World Wide Web servers are operational
1994	Yahoo is created

1995	Java is introduced
	Amazon is founded
	E-Bay is founded
	Hotmail is founded
1996	WebTV is introduced
1998	Goggle is founded
	PayPal is founded
2001	Xbox is introduced
2002	Approximately 1 billion PCs have been sold
2006	Skype has over 100 million registered users

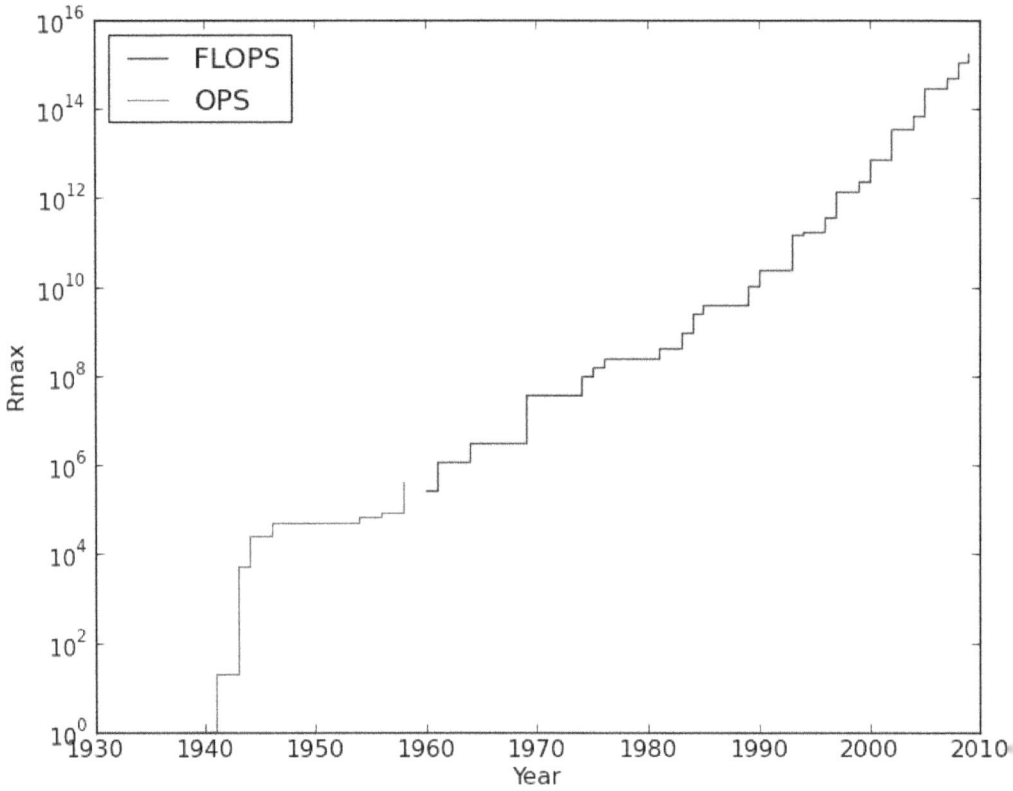

Supercomputer Speed vs. Year

Robotics

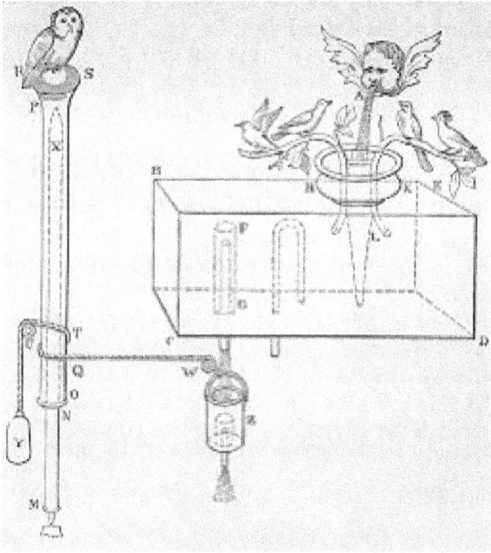

A Trumpet, in the hands of an Automaton, sounded by compressed Air – Hero
(10-70 A.D.)

Fundamentally, there are two types of robots, those that are operated remotely known as tele-robots and those which are autonomous. Spacecraft and planetary rovers are tele-robots, as humans ultimately control the actions of these devices. The autonomous variety has program control that will eventually become powerful enough to allow these devices to perform sophisticated tasks like building, cleaning, operating machinery etc. There are some crucial steps that need to occur before this point however, many of which are covered by science fiction writers like Isaac Asimov. Autonomy begets consciousness in many of his stories, this will have to be dealt with by humans. The following is a time line, albeit incomplete, that shows the major developments of robotic concepts and designs:

c. 270 BC   Ctesibius, a Greek physicist and inventor makes organs and water clocks with movable figures.

1495   The anthrobot, a mechanical man, is designed by Leonardo da Vinci.

1540   A mandolin-playing lady is created by Italian inventor Gianello Torriano.

1772   Swiss inventors Pierre and Henri Jacquet-Droz build a robotic child called L'Ecrivain (The Writer). It could write messages with up to 40 characters. L'Ecrivain's brain was a mechanical computer. A piano-playing robotic woman is also built at this time.

1801   Joseph Jacquard invents a textile machine called a programmable loom. It is operated by punch cards.

1818   Mary Shelley writes "Frankenstein" about a frightening artificial life form created by Dr. Frankenstein.

1830   American Christopher Spencer designs a cam-operated lathe.

1890's   Nikola Tesla designs the first remote control vehicles. He is also known for his invention of the radio, induction motors, Tesla coils.

1892   In the United States, Seward Babbitt designs a motorized crane with gripper to remove ingots from a furnace.

1921   The first reference to the word robot appears in a play opening in London, entitled *Rossum's Universal Robots*. The word robot comes from the Czech   word, robota, which means drudgery or slave-like labor. Czech playwright Karel Capek first used this term when describing robots that helped people with   simple, repetitive tasks. Unfortunately, when the robots in the story were used in   battle,   they   turn against their human owners and take over the world.

1938   Americans Willard Pollard and Harold Roselund design a programmable paint-   spraying mechanism for the DeVilbiss Company.

1940's   Grey Walters creates an early robot called Elsie the tortoise, or Machina speculatrix.

1941   Science fiction writer Isaac Asimov first uses the word "robotics" to describe the technology of robots and predicts the rise of a powerful robot industry.

1942   Asimov writes a story about robots, *Runaround*, which contains the "Three laws of robotics".

1946   George Devol patents a general purpose playback device for controlling machines. It uses a magnetic process recorder. American scientists J. Presper Eckert and   John   Mauchly build the first large electronic computer called the Eniac at the

University Pennsylvania. The second computer, the Whirlwind, solves a problem at M.I.T. The Whirlwind is the first general-purpose digital computer.

1948   Norbert Wiener, a professor at M.I.T., publishes his book, *Cybernetics*, which describes the concept of communications and control in electronic, mechanical,   and   biological systems.

1951   A teleoperator-equipped articulated arm is designed by Raymond Goertz for the Atomic Energy Commission.

1954   The first programmable robot is designed by George Devol. He coins the term Universal Automation.

1956   Devol and engineer Joseph Engelberger form the world's first robot company, Unimation.

1959   Computer-assisted manufacturing was demonstrated at the Servomechanisms Lab at MIT. Planet Corporation markets the first commercially available robot.

1960's Johns Hopkins creates the beast. It is controlled by hundreds of transistors and able to seek out photocell outlets when its battery runs low.

1960   The General Electric Walking Truck was a 3,000 pound, four-legged robot that could walk four miles an hour. It was powered by a computer. Ralph Moser   developed   the machine.

1960   Unimation is purchased by Condec Corporation and development of Unimate Robot Systems begins. American Machine and Foundry, later known as AMF Corporation, markets a robot, called the Versatran, designed by Harry Johnson and Veljko Milenkovic.

1961   The first industrial robot was online in a General Motors automobile factory in New Jersey. It was Devol and Engelberger's UNIMATE. It performed spot welding and extracted die castings.

1963   The first artificial robotic arm to be controlled by a computer was designed. The Rancho Arm was designed as a tool for the handicapped and its six joints gave it the flexibility of a human arm.

1964   Artificial intelligence research laboratories are opened at M.I.T., Stanford Research Institute (SRI), Stanford University, and the University of Edinburgh.

1965   DENDRAL was the first expert system or program designed to execute the accumulated knowledge of subject experts.

1968   The octopus-like Tentacle Arm was developed by Marvin Minsky.

1969   The Stanford Arm was the first electrically powered, computer-controlled robot arm.

1970   Shakey was introduced as the first mobile robot controlled by artificial intelligence. SRI International in California produced this small box on wheels that used memory to solve problems and navigate. At Stanford University a robot arm is developed which becomes a standard for research projects. The arm is electrically powered and becomes known as the Stanford Arm.

1970's	Scientists at Edinburgh University create the Freddy robot, taking steps in hand- eye coordination technology. This first assembly robot constructed a toy boat and car from a heap of mixed parts tipped onto a table.
1973	The first commercially available minicomputer-controlled industrial robot is developed by Richard Hohn for Cincinnati Milacron Corporation. The robot is called the T3, The Tomorrow Tool.
1974	A robotic arm (the Silver Arm) that performed small-parts assembly using feedback from touch and pressure sensors was designed. Professor Scheinman, the developer of the Stanford Arm, forms Vicarm Inc. to market a version of the arm for industrial applications. The new arm is controlled by a minicomputer.
1976	Robot arms are used on Viking 1 and 2 space probes. Vicarm Inc. incorporates a microcomputer into the Vicarm design.
1977	ASEA, a European robot company, offers two sizes of electric powered industrial robots. Both robots use a microcomputer controller for programming and operation. Unimation

purchases Vicarm Inc. during this year.

1978   Vicarm, Unimation creates the PUMA (Programmable Universal Machine for Assembly) robot with support from General Motors. Many research labs still use this assembly robot.

1979   The Stanford Cart crosses a chair-filled room without human assistance. The cart is equipped with a television camera mounted on a rail that takes pictures and relays them to a computer so that distances can be analyzed.

1980   The robot industry starts its rapid growth, with a new robot or company entering the market every month.

1983   The Remote Reconnaissance Vehicle became the first vehicle to enter the basement of Three Mile Island after a meltdown in March 1979. This vehicle worked four years to survey and clean up the flooded basement.

1984   The CoreSampler drilled core samples from the walls of the Three Mile Island basement to determine the depth and severity of radioactive material that soaked into the concrete.

1984   The Terregator pioneered exploration, road following and mine mapping. It was the world's first rugged, capable, autonomous outdoor navigation robot.

1985   REX was the world's first autonomous digging machine. It sensed and planned to excavate without damaging buried gas pipes. This robot used a hypersonic air knife to erode soil around pipes.

1986   The Remote Work Vehicle was developed for a broad agenda of clean-up operations like washing contaminated surfaces, removing sediments, demolishing radiated structures, applying surface treatments, and packaging and transporting materials.

1986   NavLab I pioneered high performance outdoor navigation. NavLab deployed racks of computers, laser scanners, and color cameras providing cutting-edge perception in its time.

1988   The Pipe Mapping computes magnetic and radar data over a dense grid to infer the depth and location of buried pipes. This outperforms hand-held pipe detectors.

1988   The Locomotion featured a chassis that steers and propels all wheels so that it can spin, drive, or spin while driving. Its software can emulate a tank, car or any other wheeled machine.

1990   The Ambler was a walking robot that enables energy-efficient overlapping gaits. Developed as a testbed for research in walking robots operating in rugged terrain.

1992   Neptune articulates magnetic tracks to roam the interiors of fuel storage tanks. It evaluates deterioration in floors and walls using acoustic navigation and corrosion sensing.

1992   Dante I rappels mountain sides using a spherical laser
            scanner and foot sensors. It entered the crater of Antarctica's
            Mt. Erebus but did not reach the lava lake.

1992   NavLab II was the automated HUMMER that pioneered
            trinocular vision, WARP computing, and sensor fusion to
            navigate offroad terrain.

1993   Demeter autonomously mows hay and alphalpa. It navigates
            with GPS and uses camera vision to differentiate cut and
            uncut crops.

1994   The Dante II, build by CMU Robotics, samples volcanic gases
            from the Mt. Spurr volcano in Alaska.

1997   NASA's PathFinder lands on Mars and the Sojourner rover
            robot captures images.

2000   Humanoid robots, Honda Asimo, Sony Dream Robots (SDR),
            and the Aibo robot dog are showcased.

2004   The humanoid, Robosapien is created by US robotics
            physicist and BEAM expert, Dr. Mark W Tilden.

NASA's Robonaut, used on the International Space Station

Cosmology

This timeline of cosmological theories and discoveries is a chronological record of the development of humanity's understanding of the cosmos over the last two-plus millennia. Modern cosmological ideas follow the development of the scientific discipline of physical cosmology.

ca. 16th century BC In Babylonian cosmology, particularly that depicted in the *Enuma Elis,*   the Earth and heavens were seen as a "spatial whole, even one of round   shape," revolving around the "cult-place of the deity" rather than the Earth,   and it was believed that there is a plurality of heavens and earths.

ca. 12th century BC The *Rigveda* has some cosmological hymns, particularly in the late   book 10, notably the Nasadiya Sukta which describes the origin of the universe,   originating   from   the   monistic *Hiranyagarbha* or "Golden Egg".

6th century BC The Babylonian world map shows Babylon on the Euphrates, surrounded   by a circular landmass showing Assyria, Armenia and several cities, in turn   surrounded by a "bitter river" (Oceanus), with seven islands arranged around it   so as to form a seven-pointed star. Contemporary Biblical cosmology of the   Tanakh   reflects   the same view of the Earth as a plain or a hill figured like a   hemisphere, swimming on water and overarched by the solid vault of the firmament. To this vault are fastened the stars.

4th century BC   Aristotle proposes an Earth-centered universe in which the Earth is stationary and the universe is finite in extent but infinite in time

3rd century BC   Aristarchus of Samos proposes a Sun-centered universe

2nd century BC   Seleucus of Seleucia elaborates on Aristarchus' heliocentric universe, using the phenomenon of tides to explain heliocentrism

2nd century AD   Ptolemy proposes an Earth-centered universe, with the Sun and planets revolving around the Earth

5th-11th centuries   Several astronomers propose a Sun-centered universe, including Aryabhata, Albumasar and Al-Sijzi

6th century   John Philoponus proposes a universe that is finite in time and argues against t   he ancient Greek notion of an infinite universe

ca. 8th century   Puranic Hindu cosmology, in which the Universe goes through repeated   cycles of creation, destruction and rebirth, with each cycle lasting 4.32 billion   years.

9th-12th centuries   Al-Kindi (Alkindus), Saandia Gaon (Saadia ben Joseph) and the Al-   Ghazali (Algazel) support a universe that has a finite past and develop two   logical arguments against the notion of an infinite past, one of which is later   adopted by Immanuel Kant

964   Abd al-Rahman al-Sufi (Azophi), a Persian astronomer, makes the

first recorded   observations of the Andromeda Galaxy and the Large Magellanic Cloud, the first   galaxies other than the Milky Way to be observed from Earth, in his *Book of Fixed Stars*

12th century   Fakhr al-Din al-Razi discusses Islamic cosmology, rejects Aristotle's idea of   an Earth-centered universe, and, in the context of his commentary on the   Qur'anic verse, "All praise belongs to God, Lord of the Worlds," proposes that   the universe has more than "a thousand thousand worlds beyond this world   such that each one of those worlds be bigger and more massive than this world   as well as having the like of what this world has." He argued that there exists   an infinite outer space beyond the known world, and that there could be an   infinite number of universes.

13th century   Narir al-Din al-Tusi provides the first empirical evidence for the Earth's rotation on its axis

15th century   Ali Qushji provides empirical evidence for the Earth's rotation on its axis and rejects the stationary Earth theories of Aristotle and Ptolemy

15th-16th centuries   Nilakantha Somayaji and Tycho Brahe propose a universe in which the   planets orbit the Sun and the Sun orbits the Earth, known as the Tychonic   system

1543   Nicolaus Copernicus publishes his heliocentric universe in his *De revoutionibus orbium coelestium*

Life in the Universe and Where to find it

1576   Thomas Digges modifies the Copernican system by removing its outer edge and replacing the edge with a star-filled unbounded space

1584   Giodano Bruno proposes a non-hierarchical cosmology, wherein the Copernican solar system is not the center of the universe, but rather, a relatively insignificant star system, amongst an infinite multitude of others

1610   Johannes Kepler uses the dark night sky to argue for a finite universe

1687   Sir Isaac Newton's laws describe large-scale motion throughout the universe

1720   Edmund Halley puts forth an early form of Olbers' paradox

1744   Jean-Philippe de Cheseaux puts forth an early form of Olbers' paradox

1791   Erasmus Darwin pens the first description of a cyclical expanding and contracting universe in his poem *The Economy of Vegetation*

1826   Heinrich Wilhelm Olbers puts forth Olbers' paradox

1848   Edgar Allan Poe offers first correct solution to Olbers' paradox in *Eureka: A Prose Poem*, an essay that also suggests the expansion and collapse of the   universe

1905   Alberty Einstein publishes the Special Theory of Relativity, positing that space and time are not separate continuum

1915   Albert Einstein publishes the General Theory of Relativity, showing that an energy density warps spacetime

1917   Willem de Sitter derives an isotropic static cosmology with a cosmological constant, as well as an empty expanding cosmology with a cosmological constant, termed a de Sitter universe

1920   The Shapley-Curtis Debate takes place at the Smithsonian

1921   The National Research Counsel_(NRC) published the official transcript of the Shapley-Curtis Debate

1922   Vesto Slipher summarizes his findings on the spiral nebulae's systematic redshifts

1922   Alexander Friedmann finds a solution to the Einstein field equations which suggests a general expansion of space

1924   Edwin Hubble discovers that the universe is composed of thousands of galaxies.

1927   Georges Lemaitre discusses the creation event of an expanding universe governed by the Einstein field equations. From its solutions to the Einstein equations, he predicts the distance-redshift relation.

1928   Howard Percy Robertson briefly mentions that Vesto Slipher's redshift measurements combined with brightness measurements of the same galaxies indicate a redshift-distance relation

1929   Edwin Hubble demonstrates the linear redshift-distance relation and thus shows the expansion of the universe

1933   Edward Milne names and formalizes the cosmological principle

1934   Georges Lemaitre interprets the cosmological constant as due to a vacuum energy with an unusual perfect fluid equation of state

1938   Paul Dirac suggests the large numbers hypothesis, that the gravitational constant may be small because it is decreasing slowly with time

1948   Ralph Alpher, Hans Bethe ("in absentia"), and George Gamow examine element synthesis in a rapidly expanding and cooling universe, and suggest that the elements were produced by rapid neutron capture

1948   Hermann Bondi, Thomas Gold, and Fred Hoyle propose steady state cosmologies based on the perfect cosmological principle

1948   George Gamow predicts the existence of the cosmic microwave background radiation by considering the behavior of primordial radiation in an expanding universe

1950   Fred Hoyle derisively coins the term "Big Bang".

1961   Robert Dicke argues that carbon-based life can only arise when the gravitational   force is small, because this is when burning stars exist; first use of the weak anthropic principle

1965   Hannes Alfven proposes the now-discounted concept of ambiplasma to explain baryon asymmetry.

1965   Martin Rees and Dennis Sciama analyze quasar source count data and discover that the quasar density increases with redshift.

1965   Arno Penzias and Robert Wilson, astronomers at Bell Labs discover the 2.7 K *microwave background radiation*, which earns them the 1978 Nobel Prize in   Physics. Robert Dicke, James Peebles, Peter Roll and David Todd Wilkinson interpret it as relic from the big bang.

1966   Stephen Hawking and George Ellis show that any plausible general relativistic cosmology is singular

1966   James Peebles shows that the hot Big Bang predicts the correct helium abundance

1967   Andrei Sakharov presents the requirements for baryogenesis, a baryon- antibaryon asymmetry in the universe

1967   John Bahcall, Wal Sargent and Maaren Schmidt measure the fine-structure splitting of spectral lines in 3C191 and thereby show that the fine-structure does not vary significantly with time

1968   Brandon Carter speculates that perhaps the fundamental constants of nature must lie within a restricted range to allow the emergence of life; first use of the strong anthropic principle

1969   Charles Misner formally presents the Big Bang horizon problem

1969   Robert Dicke formally presents the Big Bang flatness problem

1973   Edward Tryon proposes that the universe may be a large scale quantum mechanical vacuum fluctuation where positive mass-energy is balanced by negative gravitational potential energy

1974   Robert Wagoner, William Fowler, and Fred Hoyle show that the hot Big Bang predicts the correct deuterium and lithium abundances

1976   Alex Shlyakhter uses samarium ratios from the Oklo prehistoric natural nuclear fission reactor in Gabon to show that some laws of physics have remained unchanged for over two billion years

1977   Gary Steigman, David Schramm, and James Gunn examine the relation between the primordial helium abundance and number of neutrinos and claim that at most five lepton families can exist.

1981   Viacheslav Mukhanov and G. Chibisov propose that quantum fluctuations could lead to large scale structure in an inflationary universe

1981   Alan Guth proposes the inflationary Big Bang universe as a possible solution to the horizon and flatness problems

1990   Preliminary results from NASA's COBE mission confirm the Cosmic Microwave Background Radiation is an isotropic blackbody to an astonishing one part in 10^5 precision,   thus   eliminating   the possibility of an integrated starlight model proposed for the background by steady state enthusiasts.

1990s   Ground based cosmic microwave background experiments measure the first peak, determine that the universe is geometrically flat.

1998   Controversial evidence for the fine structure constant varying over the lifetime of the universe is first published.

1998   Adam Riess, Saul Perlmutter and others discover the cosmic acceleration in observations of Type 1a supernovae providing the first evidence for a non-zero cosmological constant.

1999   Measurements of the cosmic microwave background radiation (most notably by the BOOMERanG see Mauskopf et al., 1999, Melchiorri et al., 1999, de Bernardis et al. 2000) provide evidence for oscillations (peaks) in the anisotropy angular spectrum as expected in the standard model of   cosmological   structure   formation.   These results indicates that the geometry of the universe is flat. Together with large scale structure data, this provides complementary evidence for a non-zero cosmological constant.

2002   The Cosmic Background Imager (CBI) in Chile obtained images of the cosmic microwave background radiation with the highest angular resolution of 4 arc minutes. It also obtained the anisotropy spectrum at high-resolution not covered before up to $I \sim 3000$. It found a slight excess in power at high-resolution ($I > 2500$) not yet completely explained, the so-called "CBI-excess".

2003   NASA's WMAP obtained full-sky detailed pictures of the cosmic microwave background radiation. The image can be interpreted to indicate that the universe is 13.7 billion years old (within one percent error) and confirm that the Lambda-CDM model and the inflationary theory are correct.

2003   The Sloan Great Wall is discovered.

2004   The Degree Angular Scale Interferometer (DASI) first obtained the E-mode polarization spectrum of the cosmic microwave background radiation.

2006   The long-awaited three-year WMAP results are released, confirming previous analysis, correcting several points, and including polarization data.

Conclusion

Scientific achievement has be rapid.

Progress in Cosmology has been equally rapid.

The development of computers has been rapid and will continue to follow Moore's law, leading to the conclusion that they will exceed our capabilities in the form of robots and central information repositories in the near future.

Chapter 12 - Speculation

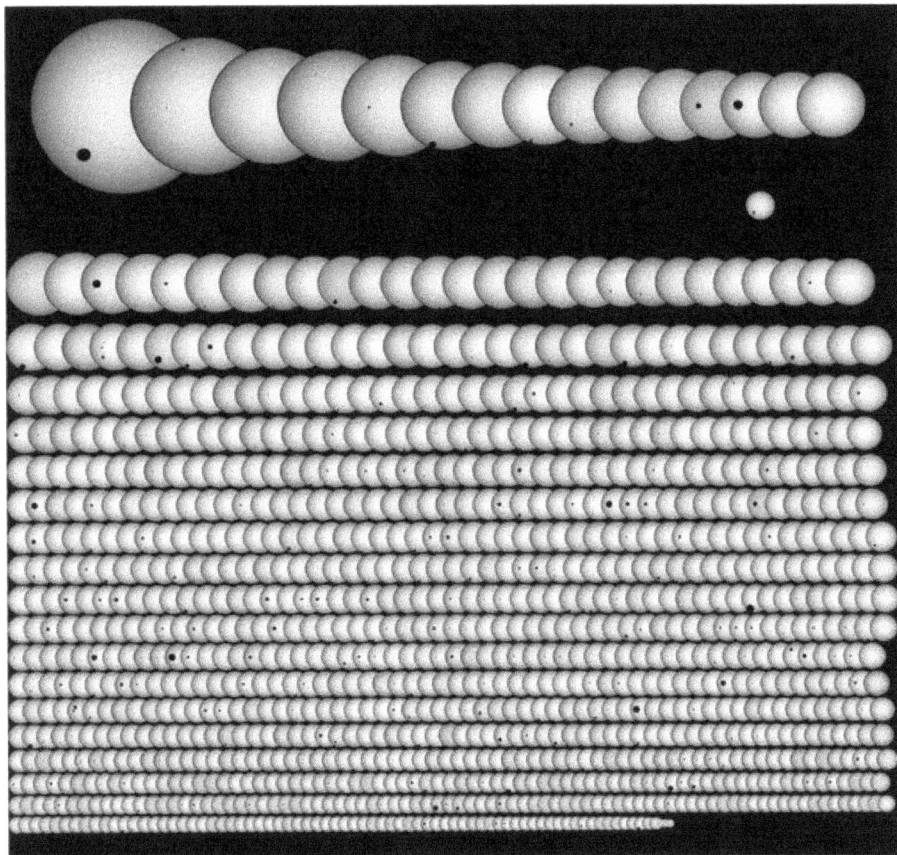

Extrasolar Stars and their Planets

Based on what we now know about Earth's life and what we have observed in the heavens, we can place boundaries on several important factors necessary for the promulgation of extra-solar life forms. In broad terms, limits have to be placed on planet size, temperature, availability of water, star type, position in orbit, atmosphere and radiation and chemistry.

One thing to consider is that the boundaries of these variables cause the propagation of life to slow down. In other words, too cold or too hot in terms of temperature is not ideal, just as too big a planet or too small.

Size of potential planets

Although this could be a subject for detailed mathematical modeling, the results of which will show that there are many variables to considers. For this treatise, we shall give only a rounded estimate for the planet size that will most easily harbor life. The estimate is based on gravity more than diameter as different compositions of planets have different gravity constants. Assuming life as Earth has produced it is estimated that

.5 Earth to 5 Earth diameters

would be a reasonable range for planetary size

Temperature range for life

Based on the fact that life on Earth has been found active at just above freezing to well above 50 C and that the vast abundance of intelligent life is found in a smaller range of

342

0C to 50C

It is estimated that this "sweet spot" would be conducive to life on other planets as well. Above and below this range, life exists of course but in a much more "labored" environment. Life can exist close to absolute zero, but it is not mobile or capable of reproduction. Consider a Gaussian curve distribution of life vs. temperature on Earth with the majority existing around 20 Degrees C. It is hard to say if this curve is universal relative to life in extra-solar planets, but analysis of the abundance of life and the distribution of chemicals in space indicate that there are similarities that allow us to extrapolate this range.

Water

Significant amounts of water have been found by radio telescopes in interstellar space. It is well known that life evolved on Earth in the presence of water. 3 Billion years ago, plants first made their appearance and required water. This followed a period of some 800 Million years were single and multiple cell life evolved out of what is colloquially called the "primordial ooze." Although there are several researchers that have argued that water is not necessarily required, for the purposes of this book, we know that life in the presence of water has worked and estimate the likelihood of extrasolar planets requiring water is:

.1 Earth to Water World

No doubt life can exist in only water and we predict that even a small amount of water will be conducive to the generation of life. All Earthly animals consist of a high percentage of water, so it is possible that a world with a paltry amount of the substance will not evolve as quickly as our world did. Also, the Urey-Miller experiment proved that the building blocks of life can be generated with water.

Star Type

It is generally agreed that most stars which have been stable for billions of years that have planets in the "Goldilocks zone" can harbor life. The interesting question is now that we know that some of the stars in the universe are very old (for instance giant type stars) and assuming that life evolved on their planets at the same time after formation and at the same rate; how advanced is life there now? Considering the limits of star temperature, we estimate that the range of star types will be:

B > K

Weight Range of animals

There is a very wide range of the size of animals and of life in general. From the smallest bacteria to the largest animal to ever exist covers magnitudes of weight range. There has been speculation about life that can exist in the air or clouds. Airborne viruses are an obvious example of this type of life, but they do not live the majority of their lives in this arena. Considering the size of the extrasolar planet it is estimated that the range of weights will be:

Aquatic

> **Microscopic to Blue whale at 1 Earth**
>
> **Microscopic to 2 Blue whale at .7 Earth**
>
> **Microscopic to .5 Blue whale at 2 Earth**
>
> **Microscopic to .1 Blue whale at 5 Earth**

Land

> **Microscopic to at 1 Amphicoelias fragillimus at 1 Earth**
>
> **Microscopic to 2 Amphicoelias fragillimus at .7 Earth**
>
> **Microscopic to .5 Amphicoelias fragillimus at 2 Earth**

Largest Land Animals

Air Pressure

Air pressure is important for the transport of life sustaining oxygen and carbon dioxide. Extrapolating what pressures exist at the deepest depths of the oceans where life has been found and this leads us to limit the range to:

.6 to 30 standard atmospheres

.6 atmosphere corresponds to 12,000 feet of altitude, which in general terms is known as the "tree line." Lichen and moss has been found at higher altitudes but the abundance is low enough to make 12,000 the general limit.

Air Chemistry

There is probably a wide variety of atmospheric chemistry that will allow the presence of life. It is beyond the scope of this book to go into the fine details of this chemistry. Instead we will estimate the amount of certain elements necessary for an Earth like planet:

Oxygen	**.5 to 40%**
Nitrogen	**30 to 80%**
Carbon Dioxide	**10 to 40%**

Again, there is a wide amount of speculation on this chemistry, including the fact that other chemicals could harbor life as well.

Radiation

It is well known that high amounts of radiation can cause cell damage in living organisms. It is also known that the Earth's atmosphere can shield significant amounts of this radiation. This includes ultra-violet, cosmic rays and other radiation caused by variations in the solar cycle in the form of flares, etc. A very general range of radiation would be:

0 to 25 Rems

The upper limit of this range can be tolerated for only a small period of time, a lower limit of about 10 can be acceptable for longer periods.

Time vs. Life with other planets unlike Earth (what could speed up or slow down progress?)

The time life starts and the rate it evolves is highly dependent on the major variables mentioned above and quite a few other minor ones. There is reason to believe that life could propagate faster under the most ideal conditions and slower under less than ideal conditions. Planets and moons such as Mars, Europa and Enceladus could have had or presently have life of some sort but considering their challenging environments it is believed that the speed of which life evolves is slowed and the diversity of life is limited. Therefore we bracket the time of the start of life as between:

100 Million and 1 Billion Years

Number of planets in habitable zone that contain:

Life of any sort (simple to complex)	**80%**
Plant Life	**70%**
Animal Life	**30%**
Sentient Life	**10%**

The basis of this speculation is upon the examination of the historical ease of how life of a particular sort was able to flourish on Earth. Multiple

347

scenarios are possible and the influence of such things as asteroid and comet impacts can change the numbers above dramatically.

The other assumption about the speculation is that the candidate planet is as old as the Earth is now.

Probability Calculation of Life

Using a Standard Deviation approach to statistically predict the probability of life on an extraterrestrial planet, we can propose the following equation:

$$Probability = SSize * STemp * SWater * SType * SWeight * SPress *$$
$$SChem * SRad * STime * SPos * SJup$$

where:

SSize = % of numbers in an optimal distribution of The Size of the Planet

STemp = % of numbers in an optimal distribution of The Average Temperature of the Planet

SWater = % of numbers in an optimal distribution of The

Amount of Water Coverage on the Planet

SType = % of numbers in an optimal distribution of The Type of Sun for the Planet

SWeight = % of numbers in an optimal distribution of The Weight of the the life on the Planet

SPress = % of numbers in an optimal distribution of The Atmospheric Pressure of the Planet

SChem = % of numbers in an optimal distribution of The Chemicals found in the atmosphere of Earth of the candidate Planet

SRad = % of numbers in an optimal distribution of The Safe levels of Radiation of the Planet

STime = % of numbers in an optimal distribution of The Time of stable environment of the Planet

SPos = % of numbers in an optimal distribution of The Position in orbit of the Planet relative to the "Goldilocks" zone

SJup = % of numbers in an optimal distribution of The amount of planetary systems with a Jupiter like planet

This equation can be applied to an "average" starfield, for instance the local neighborhood of our sun within 50 light years. Average in this case implies the lack of nova, supernova or other unstable stars. The population density is also assumed to be a few thousand stars or less.

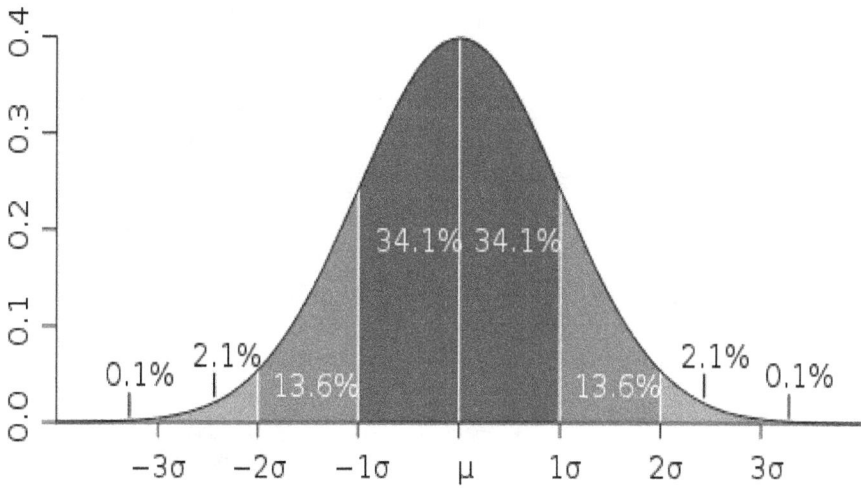

Standard Deviation of a Normal Distribution. Ultimately the probability of life on extraterrestrial planets will follow this curve, where the center is represented by the confluence of favorable conditions and history.

Chapter 13 - Conclusion

Species number

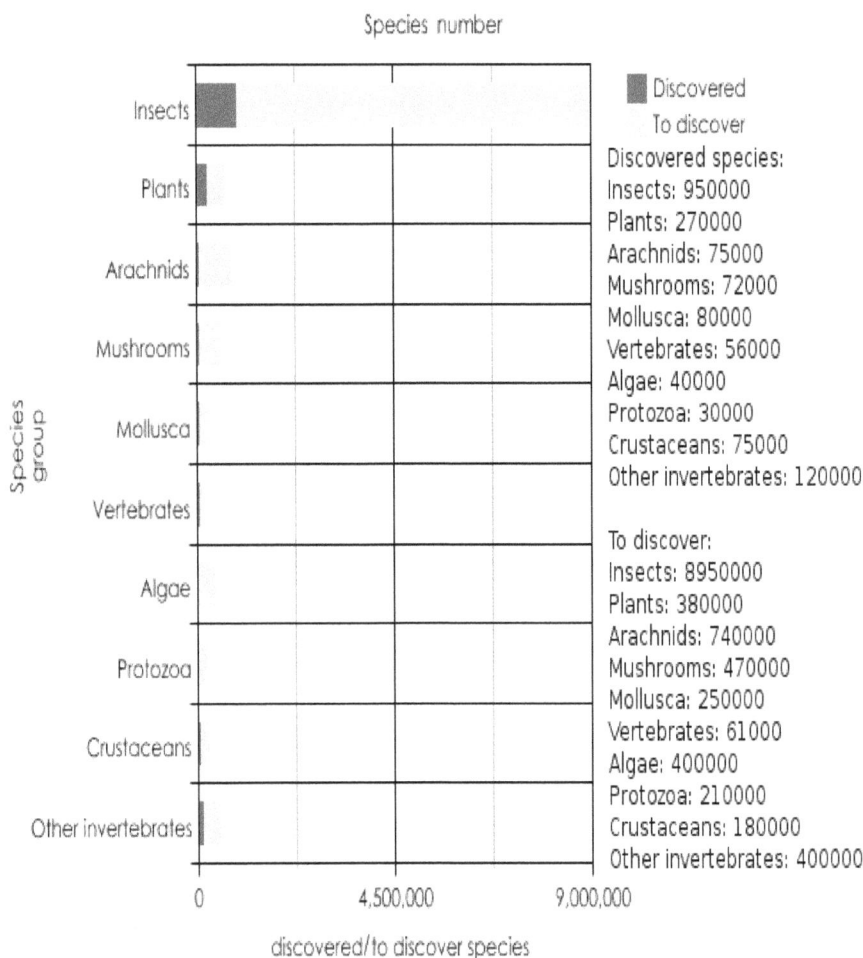

Discovered
To discover

Discovered species:
Insects: 950000
Plants: 270000
Arachnids: 75000
Mushrooms: 72000
Mollusca: 80000
Vertebrates: 56000
Algae: 40000
Protozoa: 30000
Crustaceans: 75000
Other invertebrates: 120000

To discover:
Insects: 8950000
Plants: 380000
Arachnids: 740000
Mushrooms: 470000
Mollusca: 250000
Vertebrates: 61000
Algae: 400000
Protozoa: 210000
Crustaceans: 180000
Other invertebrates: 400000

Species group

Insects
Plants
Arachnids
Mushrooms
Mollusca
Vertebrates
Algae
Protozoa
Crustaceans
Other invertebrates

0   4,500,000   9,000,000

discovered/to discover species

Number of Species so Far

Life in the Universe and Where to find it

<u>What have we learned so far?</u>

Our system worked

Given our history, chemistry and environment, life on Earth has flourished. There are and have been millions of species, covering a vast spectrum of sizes, locations and types. This includes three general classifications: plant, animal and viral (old terms)

Life is tough and resilient

Life has been found just about everywhere on Earth, cold and hot areas, under the sea, lying dormant in arid areas. It has proven easy to start and hard to kill. It has been found to be amazingly adaptable and quite possibly on Mars and other solar system objects.

Life can occur naturally or be delivered

Abiogenesis and Panspermia are the leading candidates for the origins of life on Earth. Abiogenesis shows how easily the precursors, amino acids, can be generated. The issues of course is to connect the dots from that simple place to advanced life. The model will take some time but is in process now. There is also evidence that life could have been delivered by meteor impact or possibly by cometary material. The reality is that both competing theories could be true.

352

We have not put out enough electromagnetic energy to be detected

> When one considers the power, antenna patterns, ionosphere and frequency of operation, very little if any usable modulated emissions can be detected beyond our own solar system. Using sophisticated techniques to enhance detectability assumes that some entity out in the stars has developed the same technology. This is presumptuous.

We cannot assume other close worlds are watching us are using the same technology as us

> This conclusion relates to communications as well as all other forms of technology. It is presumptuous on our part to assume all life will follow the same path. It is more accurate to say that little or no other life will follow our path. Therefore, sentient life on other planets will be different, either in a more or less advanced way.

Philosophers, scientists, cosmologists and science fiction writers have pondered the same question

> It seems to be in our nature to explore, to consider what is on the next mountain and what form life takes in the stars. This truly is a philosophic question. One that has been discussed amongst the sages as well as scientists. It is a natural consequence of intelligence

and teaches us that even intuitions can turn out to be true.

Jupiter puts out 600 Billion watts at low frequency

Like a slowly rotating pulsar, Jupiter puts out enormous amounts of power in the form of cyclotron emissions at a particular narrow band of frequencies in the High Frequency (HF) area of the electromagnetic spectrum and in a particular direction. Do other similar planets in other solar systems do the same? Probably. Considering that Jupiter puts out more power than any other planet and in a predictable way, should we not look in the cosmos for similar signatures, as the existence of such a gas giant has been proven to be advantageous for life on Earth.

Jupiter like planets are important in a solar system as filters

Watching the Shoemaker/Levy-9 comet fall into Jupiter in 1994 reminds us that Jupiter for one and the other gas giants additionally, act like giant vacuum cleaners in our solar systems. They constantly sweep up meteorites, comets and other space debris. Some of this detritus would have fallen on Earth if it was not for these planets. One needs to ask how many disasters Earth has avoided due to our giant friends? It seems advantageous to have such a solar system arrangement.

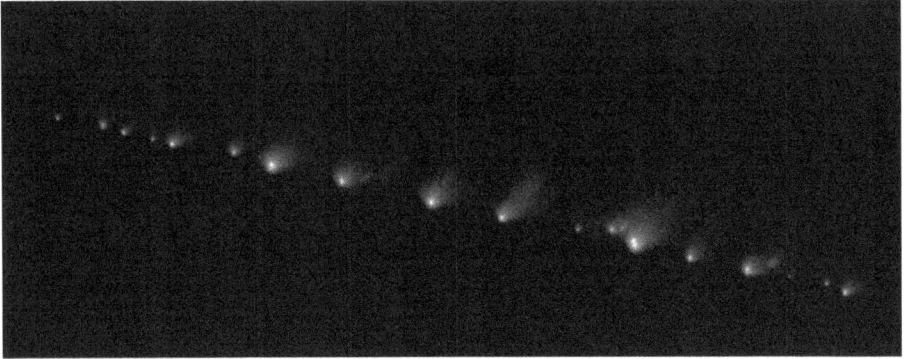

Shoemaker/Levy-9

Radio telescopes are the primary tool for both another Jupiter and chemicals

These telescopes are best suited to receive relatively lower frequencies and thus farther traveling signals. Path loss is less for the lower frequencies. Also, radio telescopes have pioneered spectroscopic measurement of the Universe and have successfully mapped regions of particular chemicals like water, carbon monoxide etc.

The distances are too far and thus too lossy to detect communications, etc.

Several simple calculations show this to be true. It has also been assumed that "they" would have very high power transmitter, very large antennas and would be pointed directly at us. This would assume that we would have very large antennas pointed directly at

355

them. Large antennas have very small beamwidths and thus the possibilities of this level of collusion is extremely remote.

Spectroscopic radio astronomy offers a solution

It is a significant advantage to be able to "dial in" a particular chemical or locate a plethora of chemicals over a particular bandwidth while looking at a stellar object. With new high resolution instruments like the Atacama array coming online, the ability to examine the chemical compositions of the atmospheres of extraterrestrial planets is a real possibility. The signatures are strong enough, especially with long term integration, to be observable. Life leaves its signature in the atmosphere so it is probably the best way to confirm its existence in the stars.

Same 92 elements found in the universe

This leads us to believe that "our" chemistry is the same as "theirs." This gives us confidence in predicting what we should observe. We should however, expect the ratios to be different and thus creating a large amount of diversity.

Radio telescopes are capable of detecting biological signs of life

These instruments are best suited for examining low frequency emissions and are capable of great sensitivity. This is true for Jupiter like radio outbursts as well as looking for the chemical signatures of life in extraterrestrial atmospheres. Integration periods of years is possible thus allowing phenomenal detection capability.

Occultation and spectral shifting have discovered other planets

These techniques have led to thousands of extraterrestrial planet discoveries in a relatively short period of time. No doubt thousands more will be discovered in the next few years. In addition, the discoveries have led to several satellites, like the Kepler observatory, that will be able to make much more finer measurements of these planets' mass and periods.

Many of the techniques used for planet discovery rely on a co-planar aspect

Spectroscopic and occultation techniques require that the star systems we are examining for evidence of planets, are co-planar. In other words the plane of the orbits of the candidate planets are in line with us, otherwise the only method we could use for planetary detection would be astrometry or movement of the star relative to the background.

Life is vast in terms of variety (size, composition, intelligence)

Life in the Universe and Where to find it

Hundreds of millions of species have been discovered or will be discovered on Earth alone. The extraordinary extent of environments, sizes and variety of life leads one to believe that it can adapt to almost any reasonable world.

Much more simple life than complex

Simple life, that will but a few cells for instance, represents approximately 99% of that discovered as compared to larger vertebrates. Sentient life represents a very small percentage of vertebrate life. Therefore, intelligent life is rare, non-sentient is plentiful. This of course, assumes that we know how to define sentience.

Timelines indicate that there is a non linear increase in capability

The timelines of human development show that paths of technological discovery take circuitous routes. All paths of development could have taken drastically different directions, teaching us that we should not assume that other life in the Universe will follow the same directions. It is therefore presumptuous to assume that other life will develop radar or interferometers but in fact could quite possibly develop technologies that we have not have even dreamt of.

Best guess

Life is abundant and close, much of it in <300 Ly

> Several thousand stars exist in this area, most have been detected with planets. Considering the defining elements conducive for life, most of the qualifying planets will have life. Most if not all of these planets will have simple life. Some of these could harbor more sophisticated forms, with but a few containing intelligent, or sentient life. Best guess is life exists on over 50% of the qualified planets, sentient life on less than 1%.

Will be detected by radio telescopes using long integration periods and spectroscopic techniques

> These instruments have essentially infinite sensitivity as they can spend hours, days and weeks staring at the same candidate planets for evidence of life. The life will initially be found by chemical elements in the atmospheres indicating their existence on the surface or ocean below.

The easiest life to be detected will be the simplest as it will be the most abundant

> By far the most abundant life will be simple, which go through chemical reactions to process their "food." This process gives off residual gases and chemicals

that can be detected by telescopes.

Within 10 years as Kepler and ESO satellites and ground based observatories are finding planets quickly

The pace of planetary discovery has intensified over the last few years. In addition, orbiting observatories have discovered many more and have even allowed the public to review data to help in reducing the massive amount of data to discover yet more planets. Thousands of candidates have been found, many hundreds have been confirmed as of this writing

Evolution is universal

The fundamental idea that the strong survive and that life changes with each succeeding generation will be found true on other planets. Thus, evolutionary tracks will be found, but none like our own as we have a unique environment and history.

Variations on our system will work, within limits

Life can exist in a wide range of environments on Earth. This will also be the case on other worlds. The fact is that other worlds will have different average temperatures, different atmospheric chemistries and different gravity constants. These and other factors will mold the life on these worlds in a unique way.

Lower forms have better chance

Simple life reproduces quickly allowing for ease of adaptation, therefore just like on Earth, these forms of life will be the first to appear and be the most prolific.

Simple life much easier to occur, much harder to find

Simple life is small by nature and can take the form of algae, lichen or fungi.

We Can't:

Assume life there will be like life here

Even with minor environmental differences, which are inevitable on other planets, life will react in ways different from our own history. Thus, the chances of identical life on interstellar planets is near impossible.

Intelligence will evolve in the same way

Intelligence on Earth has peaked with bipeds. There is no reason to believe that this is a necessary trait. Notice that many other species have the ability to interact and are possibly sentient. Examples are dolphins, pets, etc. Where will these species be in several more million years? This time period is small compared to geologic time or even the amount of time life has been on Earth.

Life in the Universe and Where to find it

Assume Philosophy, religion and other sociological details will evolve in the same way

> The human brain has spawned these areas of thought and there is no guarantee that brains from another star system will create the same streams of thought. It is in fact arrogant to assume that they will. In reality a more likely case is that sentient thought on other worlds will always take different paths.

How do you find life in outer space?

Look for Jupiter like emissions – Shoemarkian Array

> As of the publication of this book, thousands of other worlds have been discovered. Within a few years of this book's publication, many more thousands will be discovered. From this list, a significant amount of solar systems will be identified that have Earth like planets in the "Goldilocks" orbits and gas giants like Jupiter in outer orbits. These giants are capable of radio emissions like Jupiter and considering the magnitude of the radio bursts and periodicity, can be detected on Earth with dedicated low frequency radio telescopes. The best option is to use a "Shoemarkian" array where very wide angles of sky can be observed simultaneously and high detailed images can be created using post processing.

Use MRAS model (element beamwidth <u>and</u> array beamwidth) to remove phase

MRAS or the Mobile Radio Astronomy System was designed to view areas of the sky within the beamwidth of a single element, then, using interferometric techniques, observe very fine details with beam widths equivalent to that of a single antenna the breadth of the the array dimensions. This array is optimal for looking for chemical signatures at millimeter and submillimeter wavelengths.

Look for chemical signatures

The extent of chemical signature in a candidate extrasolar planet is such that there is a detectable signal over a large area. Using long term integration approaches, very low amounts of trace chemicals can be found. The very high resolution of a distributed array lends itself to discerning different planets' atmospheres from others. An array like NRAO Atacama is a prime instrument for such a search.

Design comprehensive mathematical model

To fully understand the extent of the possibilities of life and be able to manipulate the environments they exist in, a good comprehensive model needs to be built that can predict what path life takes given the environment and history of a candidate planet. In this was, when life is in fact discovered out there, the model can be updated to allow better predictions of the details.

Where to Find it?

Looking within the plane of the Milky Way gives the best chance of finding the chemical markers for life in the Galaxy, but it is also the noisiest part of the sky. Areas in the sky away from the galactic plane lower the noise level and include many of the closest star systems. The optimal solution would be to find candidate solar systems within 100 Light Years outside of the plane and concentrate our efforts there.

What are we going to do when it happens?

There was a time when fear dominated the possibility of learning about life in the Universe. Then there was the interesting speculation that if we did find life out "there" we would recognize our similarities instead of our differences, leading to a more peaceful planet. Now probably, we are just looking forward to getting the discovery behind us and proceeding to witness it ourselves, in whatever form. We are ready now.

Simple Laws:

1. Life is easy to start in a reasonable environment

2. Evolution from the standpoint of planets harboring life, will not take the same path twice

3. Most life in the Universe is not intelligent

4. Sentient thought will not take the same path twice

Appendices:

Drake

The first systematic attempt to detect artificial radio signals from nearby stars. Named after the princess in Frank Baum's *Wizard of Oz*, it was the brainchild of American radio astronomer Frank Drake working at the Green Bank observatory in West Virginia. Drake began preparations for Ozma in 1959, the same year in which the seminal theoretical paper on SETI by Philip Morrison and Guiseppe Cocconi was published in the British journal *Nature* (see Morrison-Cocconi Conjecture). These developments, although occurring more or less simultaneously (the paper appeared about 6 months after Drake began his work), were quite independent of one another. Both, however, concluded that the best chance of success would come from searching at a radio wavelength of 21.1 cm (corresponding to a frequency of 1,420 MHz) since the 21-centimeter line of neutral hydrogen in the Galaxy might represent a natural hailing wavelength at which intelligent species would try to communicate. As his target, Drake chose two nearby, reasonably Sun-like stars, epsilon Eridani and tau Ceti. From April to July 1960, he tuned into both for six hours a day, using the 85-ft. Howard E. Tatel radio telescope at Green Bank equipped with a receiver that had just a single channel and a bandwidth of only 100 Hz. Although after 150 hours of listening Ozma drew a blank, it was to be the starting point for many more, increasingly sophisticated searches which continue to this day.

Masses of animals

By considerable measure, the largest known animal on Earth is the blue whale. Mature blue whales can measure anywhere from 75 feet (23 m) to 100 feet (30.5 m) from head to tail, and can weigh as much as 150 tons (136 metric tons). That's as long as an 8- to 10-story building and as heavy as about 112 adult male giraffes! These days, most adult blue whales are only 75 to 80 feet long; whalers hunted down most of the super giants. Female blue whales generally weigh more than the males. The largest blue whale to date is a female that weighed 389,760 pounds (176,792 kg).

A blue whale's head is so wide that an entire professional football team -- about 50 people -- could stand on its tongue. Its heart is as big as a small car, and its arteries are wide enough that you could climb through them. Even baby blue whales dwarf most animals. At birth, a blue whale calf is about 25 feet (7.6 m) long and weighs more than an elephant. And they do grow up fast: During the first 7 months of its life, a blue whale drinks approximately 100 gallons (379 liters) of its mother's milk per day, putting on as much as 200 pounds (91 kg) every 24 hours. An adult blue whale can eat more than 4 tons (3.6 metric tons) of krill, a tiny shrimp-like creature, every day.

Animal Facts

Largest animal - Blue whale

Loudest animal - Blue whale

Largest land animal - African elephant

Loudest land animal (second-loudest animal) - Howler monkey

Tallest mammal - Giraffe

Fastest mammal - Cheetah

Fastest air-traveling animal – Peregrine Falcon

Smallest mammal - Bumble-bee bat

Life in the Universe and Where to find it

Longest snake - Reticulated python

Smallest animal – Ant

This puts blue whales well above any known land mammal in terms of size. Most people believe that the largest animals to ever exist on Earth were the dinosaurs. However, one of the largest land dinosaurs, the sauropod *Argentinosaurus*, weighed only about 180,000 pounds (81,647 kg). That's little more than half the size of an adult blue whale. It makes a lot of sense that the world's largest animal would be a sea creature. Land animals have to support their own weight, whereas sea creatures get some help from the water.

It is believed that at one time there were more than 200,000 blue whales. There are only about 10,000 blue whales now -- they've been on the endangered list since the mid-1960s -- and the population is not expected to recover.

Fermi's Paradox

The Fermi paradox (Fermi's paradox or Fermi-paradox) is the apparent contradiction between high estimates of the probability of the existence of extraterrestrial civilizations and the lack of evidence for, or contact with, such civilizations.

The age of the universe and its vast number of stars suggest that if the Earth is typical, extraterrestrial life should be common. In an informal discussion in 1950, the physicist Enrico Fermi questioned why, if a multitude of advanced extraterrestrial civilizations exists in the Milky Way galaxy, evidence such as spacecraft or probes is not seen. A more detailed examination of the implications of the topic began with a paper by Michael H. Hart in 1975, and it is sometimes referred to as the Fermi–Hart paradox. Other common names for the same phenomenon are *Fermi's question* ("Where are they?"), the *Fermi Problem*, the *Great Silence*, and *silentium universi* (Latin for "the silence of the universe"; the misspelling *silencium universi* is also common).

There have been attempts to resolve the Fermi paradox by locating evidence of extraterrestrial civilizations, along with proposals that such life could exist without human knowledge. Counterarguments suggest that intelligent extraterrestrial life does not exist or occurs so rarely or briefly that humans will never make contact with it.

Starting with Hart, a great deal of effort has gone into developing scientific theories about, and possible models of, extraterrestrial life, and the Fermi paradox has become a theoretical reference point in much of this work. The problem has spawned numerous scholarly works addressing it directly, while questions that relate to it have been addressed in fields as diverse as astronomy, biology, ecology, and philosophy. The emerging field of astrobiology has brought an interdisciplinary approach to the Fermi paradox and the question of extraterrestrial life.

Notes from a Recent Conference

The Global Exploration Roadmap, International Space Exploration Coordination Group, September 2011

369

-Further analysis of HD 85512 b and the other newfound exoplanets will be able to determine more about the potential existence of water on the surface.

"I think we're in for an incredibly exciting time," Kaltenegger told reporters in a briefing today (Sept. 12). "We're not just going out there to discover new continents — we're actually going out there to discover brand new worlds." [Infographic: Alien Planet HD 85512 b holds possibility of life]

The HARPS spectrograph is designed to detect tiny radial velocity signals induced by planets as small as Earth if they orbit close to their star.

Astronomers used HARPS to observe 376 sunlike stars. By studying the properties of all the alien planets detected by HARPS so far, researchers found that approximately 40 percent of stars similar to the sun is host to at least one planet that is less massive than the gas giant Saturn.

In other words, approximately 40 percent of sunlike stars have at least one low-mass planet orbiting around it. On the other hand, the majority of alien planets with a mass similar to Neptune appear to be in systems with multiple planets, researchers said.

Astronomers have previously discovered 564 confirmed alien planets, with roughly 1,200 additional candidate worlds under investigation based on data from the Kepler space observatory, according to NASA's Jet Propulsion Laboratory in Pasadena, Calif.

Despite its relative proximity to the Sun at 42 light-years, HD 40307, at an apparent magnitude of 7.17, is not visible to the naked eye. This star came within 1.97 parsecs (6.4 light-years) of the Sun about 413 thousand years ago.-

Lowest life temperature

If you mean survive if frozen, and then thaw out and come 'back to life' then the water bear, a tiny creature, has been known to survive the vacuum and almost absolute zero temperature of space and be 'revived' back to life.

A class of especially hardy microbes that live in some of the harshest Earthly environments could flourish on cold Mars and other chilly planets, according to a research team of astronomers and microbiologists.

In a two-year laboratory study, the researchers discovered that some cold-adapted microorganisms not only survived but reproduced at 30 degrees Fahrenheit, just below the freezing point of water. The microbes also developed a defense mechanism that protected them from cold temperatures. The researchers are members of a unique collaboration of astronomers from the Space Telescope Science Institute and microbiologists from the University of Maryland Biotechnology Institute's Center of Marine Biotechnology in Baltimore, Md. Their results appear on the International Journal of Astrobiology website.

"The low temperature limit for life is particularly important since, in both the solar system and the Milky Way Galaxy, cold environments are much more common than hot environments," said Neill Reid, an astronomer at the Space Telescope Science Institute and leader of the research team. "Our results show that the lowest temperatures at which these organisms can thrive fall within the temperature range experienced on present-day Mars, and could permit survival and growth, particularly beneath Mars's surface. This could expand the realm of the habitable zone, the area in which life could exist, to colder Mars-like planets."

Most stars in our galaxy are cooler than our Sun. The zone around these stars that is suitable for Earth-like temperatures would be smaller and narrower than the so-called habitable zone around our Sun. Therefore, the majority of planets would likely be colder than Earth.

In their study, the scientists tested the coldest temperature limits for two types of one-cell organisms: halophiles and methanogens. They are among a group of microbes collectively called extremophiles, so-named because they live in hot springs, acidic fields, salty lakes, and polar ice caps under conditions that would kill humans, animals, and plants. Halophiles flourish in salty water, such as the Great Salt Lake, and have DNA repair systems to protect them from extremely high radiation doses. Methanogens are capable of growth on simple compounds like hydrogen and carbon dioxide for energy and can turn their waste into methane.

The halophiles and methanogens used in the experiments are from Antarctic lakes. In the laboratory, the halophiles displayed significant growth to 30 degrees Fahrenheit (minus 1 degree Celsius). The methanogens were active to 28 degrees Fahrenheit (minus 2 degrees Celsius).

CU-Boulder Professor Diane McKnight and her research colleagues used sandbags to divert water into a stream bed in Antarctica that had been dry for 20 years. Dormant bacterial mats popped up within 24 hours.

"We have extended the lower temperature limits for these species by several degrees," said Shiladitya DasSarma, a professor and a leader of the team at the Center of Marine Biotechnology, University of Maryland Biotechnology Institute. "We had a limited amount of time to grow the organisms in culture, on the order of months. If we could extend the growth time, I think we could lower the temperatures at which they can survive even more. The brine culture in which they grow in the laboratory can remain in liquid form to minus 18 degrees Fahrenheit (minus 28 degrees Celsius), so the potential is there for significantly lower growth temperatures."

The scientists also were surprised to find that the halophiles and methanogens protected themselves from frigid temperatures. Some arctic bacteria show similar behavior.

"These organisms are highly adaptable, and at low temperatures they formed cellular aggregates," DasSarma explained. "This was a striking result, which suggests that cells may 'stick together' when temperatures become too cold for growth, providing ways of survival as a population. This is the first detection of this phenomenon in Antarctic species of extremophiles at cold temperatures."

The scientists selected these extremophiles for the laboratory study because they are potentially relevant to life on cold, dry Mars. Halophiles could thrive in salty water underneath Mars's surface, which can remain liquid at temperatures well below 32 degrees Fahrenheit (0 degrees Celsius). Methanogens could survive on a planet without oxygen, such as Mars. In fact, some scientists have proposed that methanogens produced the methane detected in Mars's atmosphere.

"This finding demonstrates that rigorous scientific studies on known extremophiles on Earth can provide clues to how life may survive elsewhere in the universe," DasSarma said.

The researchers next plan to map the complete genetic blueprint for each extremophile. By inventorying all of the genes, scientists will be able to determine the functions of each gene, such as pinpointing the genes that protect an organism from the cold.

Tree of life, divided between three major cell types, those with and without a nucleus (Bacterial Prokaryotes and Animal Eukaryotes), preceded by the root of the tree, Archea.

Many extremophiles are evolutionary relics called Archaea, which may have been among the first homesteaders on Earth 3.5 billion years ago. These robust extremophiles may be able to survive in many places in the universe, including some of the roughly 200 worlds around stars outside our solar system that astronomers have found over the past decade. These planets are in a wide range of environments, from so-called 'hot Jupiters,' which orbit close to their stars and where temperatures exceed 1,800 degrees Fahrenheit (1,000 degrees Celsius), to gas giants in Jupiter-like orbits, where temperatures are around minus 238 degrees Fahrenheit (minus 150 degrees Celsius).

The discovery of planets with huge temperature disparities has scientists wondering what environments could be hospitable to life. A key factor in an organism's survival is determining the upper and lower temperature limits at which it can live.

Although Martian weather conditions are extreme, the planet does share some similarities with the most extreme cold regions of Earth, such as Antarctica. Long regarded as essentially barren of life, recent investigations of Antarctic environments have revealed considerable microbial activity.

"The Archaea and bacteria that have adapted to these extreme conditions are some of the best candidates for terrestrial analogues of potential extraterrestrial life; understanding their adaptive strategy, and its limitations, will provide deeper insight into fundamental constraints on the range of hospitable environments," DasSarma said.

Jupiter Radio Emissions

The Jovian decametric radio emission was discovered in 1955 by B.F. Burke and K.L. Franklin at the frequency of 22.2 MHz. The emission has an upper cutoff frequency of 39.5 MHz. It can be detected from ground-based stations from the upper cutoff frequency of the emission down to the cutoff frequency of the terrestrial ionosphere which is usually around 5 to 10 MHz. The peak of the intensity of the emission occurs at around 8 MHz. The emission occurs in episodes called "storms". A storm can last from a few minutes to several hours.

Two distinctive types of bursts can be received during a storm. The L bursts (L for Long) are bursts that vary slowly in intensity with time. They last from a few seconds to several tens of seconds and have instantaneous bandwidth of a few MHz. The S bursts (S for Short) are very short in duration, have instantaneous bandwidth of a few KHz to a few tens of KHz, and drift downward in frequency at a rate of typically -20 MHz/sec. They arrive at rates from a few to several hundred bursts per second. In a 5 KHz bandwidth receiver they last for only a few milliseconds. Sometimes both types of bursts can be heard simultaneously.

In order to study the wide band structure of the Jovian decametric radio emission, the University of Florida operates a broadband radio spectrograph. The spectrograph covers the frequency range from 18 to 36 MHz and uses the conical log spiral array. Please click on the link below for examples of wide band dynamic spectra of Jovian decametric radio emission.

The Jovian decametric radio emission is always received against the galactic background radio emission. The galactic background radio emission is generated by relativistic electrons spiraling in the weak galactic magnetic field.

The probabilities of detecting the emission depend strongly on the values of the Jovian central meridian longitude (CML), the Io Phase, and the Jovicentric declination of the Earth (DE). The CML is the value of the System III longitude of Jupiter facing the Earth. The Io Phase is the angle of Io, one of Jupiter's moons, with respect to superior geocentric conjunction. The combination of CML and Io phase values that have increased probabilities of emission are called sources. The sources are named Io-A, Io-B, and Io-C for the Io-controlled emission and A, B, and C for the Non-Io controlled emission.

Characteristics of emission, Io Related sources

Life in the Universe and Where to find it

SourceCML (degrees)Io Phase (degrees):

Io-A200-290195-265 Right Hand Circularly polarized,

L burstsIo-B90-20075-105 Right Hand Circularly polarized,

S burstsIo-C290-10225-250 Left Hand Circularly polarized L and S bursts

Non-Io Related Sources

A200-290B90-200C290-10

The emission is usually either right (RH) or left hand (LH) circularly or elliptically polarized, depending on the source.

Definitions of terms

Blackbody

A body that absorbs all the radiation falling on it. It has no reflecting power. It is also a perfect emitter of radiation. The concept of a blackbody is hypothetical. However, the radiation from stars can be described by assuming that they are black bodies. Black-body radiation is the thermal radiation that would be emitted by a black body at a particular temperature.

Cyclotron Maser Instability [10]

Life in the Universe and Where to find it

Five planets, Earth, Jupiter, Saturn, Uranus, and Neptune, produce powerful radio emissions via a mechanism known as the cyclotron maser instability. These intense radio emissions from the five planets have several common characteristics:

The radiation is associated with a planetary magnetic field;

The emission occurs at a frequency very close to the electron cyclotron frequency,

$$f_c = (1/2p)eB / m_e ,$$

where e is the electron charge, B is the magnetic field strength, and m_e is the electron mass.

The polarization of the emitted radiation is primarily right-handed polarized with respect to the magnetic field in the source region.

The emission usually occurs in regions of very low plasma density.

These characteristics are all consistent with a plasma instability called the cyclotron maser instability. For this instability an electromagnetic wave acts to organize the phase of the cyclotron motion in such a way that the electrons radiate in phase with the original wave.

Chromosphere
The lower part of the Sun's outer atmosphere that lies directly above the Sun's visible surface (photosphere)

Life in the Universe and Where to find it

Dynamic Spectra

A plot of the intensity of a signal as a function of frequency and time

Electromagnetic radiation

A flow of energy that is produced when electrical charged bodies, such as electrons, are accelerated

Io Phase

The orbital position of Io can be defined by something called the Io phase. The Io phase is 0 degrees when Io is directly behind Jupiter as seen from Earth. The Io phase increases as Io orbits until it becomes 180 degrees when Io crosses in front of Jupiter as seen from Earth.

Io Plasma Torus

The Io plasma torus is a ring of plasma surrounding Jupiter near the orbit of Jupiter's moon Io. The plasma consists mostly of oxygen and sulfur ions which originate from Io and are caught up in Jupiter's magnetic field. As the magnetic field rotates with Jupiter it sweeps the plasma into a torus around Jupiter. The ions emit radiation mostly in the extreme ultraviolet, but some emissions are observable by earth-based telescopes.

Jovian Decimetric and Decametric Radio Emission

Jupiter not only emits thermal radiation but also non-thermal radiation in two frequency bands: Decimetric radiation between 1000 and 3000 MHz

379

and the decametric radiation between 5 and 40 MHz. Radiation below 5MHz is absorbed by the Earth's atmosphere and cannot be absorbed on the ground.

Jovian central meridian longitude (CML)

CML is defined by the longitude of Jupiter facing the Earth at a certain time.

Modulation lanes

A spectral feature present in the dynamic spectra of the Jovian decametric radio emission. Modulation lanes were reported by J. Riihimaa in 1968. They consist of drifting curved structures

Non-thermal radiation

Electromagnetic radiation, such as synchrotron emission, that is produced by the acceleration of electrons or other charged particles but is non-thermal in origin, i.e. its spectrum is not that of a perfect black-body radiator.

Polarization

A measure of the way in which light or other electromagnetic radiation from a celestial body is affected by factors such as scattering due to cosmic dust or strong stellar or interstellar magnetic fields, or reflection from a surface. It can also be described as the degree to which the direction of the electric or magnetic vector in an electromagnetic wave changes in a regular fashion. Waves in which the electric vectors are entirely vertical or horizontal with respect to the direction of motion are described *plane or linearly polarized.* In general, both polarizations are present and the wave is then elliptically polarized in the right handed or left-handed sense accordingly as the resultant vector rotates clockwise or anticlockwise when viewed along the direction of motion of the wave.

Radio Spectrograph

An instrument used in radio astronomy for obtaining the intensity of the emission as a function of frequency.

Solar Bursts (Type III)

The Type III bursts are short, strong bursts that begin right after a visible flare and move rapidly from around 500MHz to lower frequencies. At low frequencies (less than 25 MHz) the drift is very slow. The bursts display a slow negative drift rate of 1MHz/sec. These bursts are caused by solar flares which eject high energy electrons into space zipping away from the sun at about 1/4 the speed of light. These electrons excite radio waves in space as they move along in the Sun's outer atmosphere. Since the density of the plasma in space falls off as you move away from the sun, these zipping electrons cause radio emission at successively lower frequencies.

Synchrotron radiation

Electromagnetic radiation from very high energy electrons moving in a magnetic field. It is an example of non-thermal emission.

Thermal Radiation

Electromagnetic radiation resulting from interactions between electrons and atoms or molecules in a hot dense medium.

Observing the Jovian Decametric Emission.

The best conditions for detecting the emission occur during the early morning hours, between midnight and 7 AM, when the terrestrial ionosphere becomes more transparent thus reducing the amount of interference. During the summer months in the northern hemisphere (June-August), the amount of interference due to terrestrial lightning storms worsen the conditions for detecting the emission. Lightning discharges produce a lot of electrical noise which is picked up by low frequency antennas either from direct propagation or by reflection in the ionosphere . The best observing conditions for detecting the emission are from October to April, provided that the planet is visible during these months.

Antennas and receivers for detecting the emission.

The Jovian decametric emission can be detected with simple antennas such as a half-wavelength dipole or other low gain antennas such as the long-wire and full wavelength loop antennas. However, such low gain antennas may allow the detection of some of the strongest bursts only. An array of two half-wavelength dipoles, separated by about one half wavelength can provide enough gain to detect most of the storms and is relatively simple to build. Antennas with gains of 6-10 dB with respect to a half wavelength dipole are more suitable for the systematic detection and study of the emission. Five-element Yagi and dipole log periodic antennas usually have gains in this range and have been traditionally used for this purpose. These higher gain antennas connected to a short wave amateur radio receiver can easily detect most of the strong part of the Jovian decametric radio emission. All these antennas, including the simple half- wavelength dipole are directional. Jupiter must be within the beam of the antenna in order to receive the emission.

Most amateur short wave radio receivers have relatively narrow passbands and adequate noise figure and can be used for detecting the emission. The relative narrow band (usually around 6 KHz) of these receivers will help in tuning away from radio stations. The receiver noise is usually just a fraction of the galactic background noise temperature. In order not to compress the output of the receiver which will limit the ability of detecting the emission, it is necessary to turn off or disable the AGC (automatic gain control) of the receiver.

An observing frequency between 18-24 MHz is recommended. At frequencies below 18 MHz strong interference from stations and lightning discharges is expected, which reduces the chances of detecting the emission. At frequencies higher than 24 MHz, the probabilities of detection drop sharply because of the drop in intensity of the emission.

A technical note:

The intensity of the emission can be more specifically expressed in terms of the parameter called flux density (power per unit area per unit bandwidth). As a reference, the minimum detectable flux density expected for an 8 dB gain linearly polarized antenna connected to a receiver having a 5 KHz bandwidth and a post detection time constant of 1 second is of the order of 5×10^{-22} wm^{-2} Hz^{-1} at a frequency of 18 MHz. Jupiter radio emissions are always received against the galactic background noise. The relatively high level of the galactic background noise is what limits the sensitivity of the system. Jovian decametric radio emission with peak flux densities in the range of 10-100x10^{-22} wm^{-2} Hz^{-1} are common. Expressing the flux density in Jansky (Jy), a unit more commonly used in radio astronomy, these peak flux densities are 100,000 to 1,000,000 Jy (1 Jy= 1x10^{-26} wm^{-2} Hz^{-1}). In terms of power and voltage at the input of a receiver, 10x10^{-22} wm^{-2} Hz^{-1} is equivalent to a power of 1x10^{-9} microwatt or 0.23 microvolt over 50 Ohms. [1]

Geologic History

Hadean Eon

3800 Ma and earlier.

Date	Event
4600 Ma	The planet Earth forms from the accretion disc revolving around the young Sun.
4500 Ma	According to the giant impact hypothesis the moon is formed when the planet Earth and the planet Theia collide, sending a very large number of moonlets into orbit around the young Earth which eventually coalesce to form the Moon. The gravitational pull of the new Moon stabilizes the Earth's fluctuating axis of rotation and sets up the conditions in which life formed.
4100 Ma	The surface of the Earth cools enough for

	the crust to solidify. The atmosphere and the oceans form. PAH infall and iron sulfide synthesis along deep ocean platelet boundaries, may have led to the RNA world of competing organic compounds.
Between 4500 and 3500 Ma	The earliest life appears, possibly derived from self-reproducing RNA molecules. The replication of these organisms requires resources like energy, space, and smaller building blocks, which soon become limited, resulting in competition, with natural selection favoring those molecules which are more efficient at replication. DNA molecules then take over as the main replicators and these archaic genomes soon develop inside enclosing membranes which provide a stable physical and chemical environment conducive to their replication: proto-cells.
3900 Ma	Late Heavy Bombardment: peak rate of impact upon the inner planets by meteoroids. This constant disturbance may have obliterated any life that had evolved to that point, or possibly not, as some early microbes could have survived in hydrothermal vents below the Earth's surface; or life might have been transported to Earth by a meteoroid.

Somewhere between 3900 and 2500 Ma	Cells resembling prokaryotes appear. These first organisms are chemoautotrophs: they use carbon dioxide as a carbon source and oxidize inorganic materials to extract energy. Later, prokaryotes evolve glycolysis, a set of chemical reactions that free the energy of organic molecules such as glucose and store it in the chemical bonds of ATP. Glycolysis (and ATP) continue to be used in almost all organisms, unchanged, to this day.

Archean Eon

3800 Ma – 2500 Ma

Date	Event
3500 Ma	Lifetime of the last universal ancestor; the split between bacteria and archaea occurs. Bacteria develop primitive forms of photosynthesis which at first do not produce oxygen. These organisms generate ATP by exploiting a proton gradient, a mechanism still used in virtually all organisms.
3000 Ma	Photosynthesizing cyanobacteria evolve; they use water as a reducing agent, thereby producing oxygen as waste

	product. More recent research, however, suggests a later time of 2700 Ma. The oxygen initially oxidizes dissolved iron in the oceans, creating iron ore. The oxygen concentration in the atmosphere slowly rises, acting as a poison for many bacteria. The Moon is still very close to Earth and causes tides 1,000 feet (305 m) high. The Earth is continually wracked by hurricane-force winds. These extreme mixing influences are thought to stimulate evolutionary processes.
2700 Ma	Timeframe of cyanobacteria evolution suggested by more recent research.

Proterozoic Eon

2500 Ma – 542 Ma

Date	Event
2400 Ma	Earliest cyanobacteria
2000 Ma	First acritarchs
By 1850 Ma	Eukaryotic cells appear. Eukaryotes contain membrane-bound organelles with diverse functions, probably derived from prokaryotes engulfing each other via phagocytosis. (See Endosymbiosis)
By 1200 Ma	Sexual reproduction first appears, increasing the rate of evolution.
1200 Ma	Simple multicellular organisms evolve, mostly consisting of cell colonies of

	limited complexity. First multicellular red algae evolve
1100 Ma	Earliest dinoflagellates
1000 Ma	First vaucherian algae (ex: *Palaeovaucheria*)
750 Ma	First protozoa (ex: *Melanocyrillium*)
850–630 Ma	A global glaciation may have occurred. Opinion is divided on whether it increased or decreased biodiversity or the rate of evolution.
580–542 Ma	The Ediacaran biota represent the first large, complex multicellular organisms - although their affinities remain a subject of debate.
580–500 Ma	Most modern phyla of animals begin to appear in the fossil record during the Cambrian explosion.
580–540 Ma	The accumulation of atmospheric oxygen allows the formation of an ozone layer. This blocks ultraviolet radiation, permitting the colonization of the land.
560 Ma	Earliest fungi
550 Ma	First fossil evidence for ctenophora (comb-jellies), porifera (sponges), and anthozoa (corals & anemones)

Phanerozoic Eon

542 Ma – present

The Phanerozoic Eon, literally the "period of well-displayed life", marks the appearance in the fossil record of abundant, shell-forming and/or trace-making organisms. It is subdivided into three eras, the Paleozoic, Mesozoic and Cenozoic, which are divided by major mass extinctions.

Paleozoic Era

542 Ma – 251.0 Ma

Date	Event
535 Ma	Major diversification of living things in the oceans: chordates, arthropods (e.g. trilobites, crustaceans), echinoderms, mollusks, brachiopods, foraminifers and radioarians, etc.
530 Ma	The first known footprints on land date to 530 Ma, indicating that early animal explorations may have predated the development of terrestrial plants.
525 Ma	Earliest graptolites.
510 Ma	First cephalopods (Nautiloids) and chitons.
505 Ma	Fossilization of the Burgess Shale.
485 Ma	First vertebrates with true bones (jawless fishes).
450 Ma	Land arthropod burrows (millipedes) appear, along with the first complete conodants and echinoids.
440 Ma	First agnathan fishes: Heterostaci, Galeaspida, and Pituriaspida.
434 Ma	The first primitive plants move onto land, having evolved from green algae living along the edges of lakes. They are accompanied by fungi, which may have

	aided the colonization of land through symbiosis.
420 Ma	Earliest ray-finned fishes, trigonotarbid arachnids, and land scorpions.
410 Ma	First signs of teeth in fish. Earliest nautiid nautiloids, lycophytes, and trimerophytes.
395 Ma	First lichens, stoneworts. Earliest harvestman, mites, hexapods (springtails) and ammonoids. The first known tetrapod tracks on land.
363 Ma	By the start of the Carboniferous Period, the Earth begins to be recognizable. Insects roamed the land and would soon take to the skies; sharks swam the oceans as top predators, and vegetation covered the land, with seed-bearing plants and forests soon to flourish. Four-limbed tetrapods gradually gain adaptations which will help them occupy a terrestrial life-habit.
360 Ma	First crabs and ferns. Land flora dominated by seed ferns.
350 Ma	First large sharks, ratfishes, and hagfish.
340 Ma	Diversification of amphibians.
330 Ma	First amniote vertebrates (*Paleothyris*).
320 Ma	Synapsids separate from sauropids (reptiles) in late Carboniferous.
305 Ma	Earliest diapsid reptiles (e.g. *Petrolacosaurus*).
280 Ma	Earliest beetles, seed plants and conifers diversify while lepidodendrids and

	sphenopsids decrease. Terrestrial temnospondyl amphibians and pelycosaurs (e.g. *Dimetrodon*) diversify in species.
275 Ma	Therapsids separate from synapsids.
251.4 Ma	The Permian-Triassic extinction event eliminates over 90-95% of marine species. Terrestrial organisms were not as seriously affected as the marine biota. This "clearing of the slate" may have led to an ensuing diversification, but life on land took 30M years to completely recover.

Mesozoic Era

Date	Event
From 251.4 Ma	The Mesozoic Marine Revolution begins: increasingly well-adapted and diverse predators pressurize sessile marine groups; the "balance of power" in the oceans shifts dramatically as some groups of prey adapt more rapidly and effectively than others.
245 Ma	Earliest ichthyosaurs.
240 Ma	Increase in diversity of gomphodont cynodonts_ and rhynchosaurs.
225 Ma	Earliest dinosaurs (prosauropods), first cardiid bivalves, diversity in cycads, bennettitaleans, and conifers. First teleost fishes.

220 Ma	
	Eoraptor, among the earliest dinosaurs, appeared in the fossil record 230 million years ago. Gymnosperm forests dominate the land; herbivores grow to huge sizes in order to accommodate the large guts necessary to digest the nutrient-poor plants, first flies and turtles (*Odontochelys*). First Coelophysoid dinosaurs
215 Ma	First mammals (e.g. *Eozostrodon*), minor vertebrate extinctions occur
200 Ma	The first accepted evidence for viruses (at least, the group Geminiviridae) exists. Viruses are still poorly understood and may have arisen before "life" itself, or may be a more recent phenomenon. Major extinctions in terrestrial vertebrates and large amphibians. Earliest examples of Ankylosaurian dinosaurs
195 Ma	First pterosaurs with specialized feeding (*Doygnathus*). First sauropod dinosaurs. Diversification in small, ornithischian dinosaurs; heterodontosaurs, fabrosaurids, and scelidosaurids.
190 Ma	Pilosaurs appear in the fossil record. First lepidopteran insects (Archaeolepsis), hermit crabs, modern starfish, irregular

	echinoids, corbulid bivalves, and tubulipore bryozoans. Extensive development of sponge reefs.
176 Ma	First members of the Stegosauria group of dinosaurs
170 Ma	Earliest salamanders, newts, cryptoclidid & elasmosaurid plesiosaurs, and cladotherian mammals. Cynodonts become extinct while sauropod dinosaurs diversify.
165 Ma	First rays and glycymeridid bivalves .
161 Ma	Ceratopsian dinosaurs appear in the fossil record (*Yinlong*)
155 Ma	First blood-sucking insects (ceratopogonids), rudist bivalves, and cheilosome bryozoans. *Archaeopteryx*, a possible ancestor to the birds, appears in the fossil record, along with triconodontid and symmetrodont mammals. Diversity in stegosaurian and theropod dinosaurs.
130 Ma	The rise of the Angiosperms: These flowering plants boast structures that attract insects and other animals to spread pollen. This innovation causes a major burst of animal evolution through co-evolution. First freshwater pelomedusid turtles.
120 Ma	Oldest fossils of heterokonts, including both marine diatoms and silicoflagellates.
115 Ma	First monotreme mammals.
110 Ma	First hesperornithes, toothed diving birds.

	Earliest limopsid, verticordiid, and thyasirid bivalves.
106 Ma	Spinosaurus, the largest theropod dinosaur, appears in the fossil record.
100 Ma	Earliest bees.
90 Ma	Extinction of ichthyosaurs. Earliest snakes and nuculanid bivalves. Large diversification in angiosperms: magnoliids, rosids, hamamelidids, monocots and ginger. Earliest examples of ticks.
80 Ma	First ants.
70 Ma	Multituberculate mammals increase in diversity. First yoldiid bivalves.
68 Ma	*Tyrannosaurus,* the largest terrestrial predator of North America appears in the fossil record. First species of *Triceratops.*

CenozoicEra

65.5 Ma – present

Date	Event
65.5 Ma	The Cretaceous-Tertiary extinction event eradicates about half of all animal species, including mosasaurs, pterosaurs, plesiosaurs, ammonites, belemnites, rudist and inoceramid bivalves, most planktic foraminifers, and all of the dinosaurs excluding their descendants the birds
From 65 Ma	Rapid dominance of conifers and ginkgos

	in high latitudes, along with mammals becoming the dominant species. First psammobiid bivalves. Rapid diversification in ants.
63 Ma	Evolution of the creodonts, an important group of carnivorous mammals.
60 Ma	Diversification of large, flightless birds. Earliest true primates, along with the first semelid bivalves, edentates, carnivores and lipotyphian mammals, and owls. The ancestors of the carnivorous mammals (miacids) were alive.
56 Ma	*Gastornis,* a large, flightless bird appears in the fossil record, becoming an apex predator at the time.
55 Ma	Modern bird groups diversify (first song birds, parrots, loons, swifts, woodpeckers), first whale (*Himalayacetus*), earliest rodents, lagomorphs, armadillos, appearance of sirenians, proboscideans, perissodactyl and artiodactyl mammals in the fossil record. Angiosperms diversify. The ancestor (according to theory) of the species in *Carcharodon*, the early mako shark *Isurus hastalis*, is alive.
52 Ma	First bats appear (*Onychonycteris*).
50 Ma	Peak diversity of dinoflagellates and nanofossils, increase in diversity of anomalodesmatan and heteroconch

395

	bivalves, brontotheres, tapirs, rhinoceroses, and camels appear in the fossil record, diversification of primates.
40 Ma	Modern type butterflies and moths appear. Extinction of *Gastomis*. *Basilosaurus*, one of the first of the giant whales, appeared in the fossil record.
37 Ma	First Nimravid carnivores ("False Saber-toothed Cats") - these species are unrelated to modern-type felines
35 Ma	Grasses evolve from among the angiosperms; grasslands begin to expand. Slight increase in diversity of cold-tolerant ostracods and foraminifers, along with major extinctions of gastropods, reptiles and amphibians. Many modern mammal groups begin to appear: first glyptodonts, ground sloths, dogs, peccaries and eagles and hawks.. Diversity in toothed and baleen whales.
33 Ma	Evolution of the thylacinid marsupials (*Badjcinus*).
30 Ma	First balanids and eucalypts, extinction of embrithopod and brontothere mammals, earliest pigs and cats.
28 Ma	*Paraceratherium* appears in the fossil record, the largest terrestrial mammal that ever lived.
25 Ma	First deer.
20 Ma	First giraffes and giant anteaters, increase

	in bird diversity.
15 Ma	*Mammut* appears in the fossil record, first bovids and kangaroos, diversity in Australian megafauna.
10 Ma	Grasslands and savannas are established, diversity in insects, especially ants and termites, horses increase in body size and develop high-crowned teeth, major diversification in grassland mammals and snakes.
6.5 Ma	First hominin (*Sahelanthropus*).
6 Ma	Australopithecines diversify (Orrorin, Ardipithecus)
5 Ma	First tree sloths and hippopotami, diversification of grazing herbivores, large carnivorous mammals, burrowing rodents, kangaroos, birds, and small carnivores, vultures increase in size, decrease in the number of perissodactyl mammals. Extinction of Nimravid carnivores
4.8 Ma	Mammoths appear in the fossil record.
4 Ma	Evolution of *Australopithecus, Stupendemys* appears in the fossil record as the largest freshwater turtle.
3 Ma	The Great American Interchanges, where various land and freshwater faunas migrated between North and South America. Armadillos, opossums, hummingbirds and vampire bats traveled to North America while horses, tapirs,

	saber-toothed cats, and deer entered South America. The first short-faced bears (*Arctodus*) appear.
2.7 Ma	Evolution of *Paranthropus*
2.5 Ma	The earliest species of *Smilodon* evolve
2 Ma	First members of the genus *Homo* appear in the fossil record. Diversification of conifers in high latitudes. The eventual ancestor of cattle, *Bos primigenius* evolves in India
1.7 Ma	Extinction of australopithecines.
1.2 Ma	Evolution of *Homo antecessor*. The last members of *Paranthropus* die out.
600 ka	Evolution of *Homo heidelbergensis*
350 ka	Evolution of Neanerthals
300 ka	*Gigantopithecus*, a giant relative of the orangutan dies out from Asia
200 ka	Anatomically modern humans appear in Africa. Around 50,000 years before present they start colonising the other continents, replacing the Neanderthals in Europe and other hominids in Asia.
40 ka	The last of the giant monitor lizards (*Megalania*) die out
30 ka	Extinction of Neanderthals
15 ka	The last Woolly rhinoceros (*Coelodonta*) are believed to have gone extinct
11 ka	The giant short-faced bears (*Arctodus*) vanish from North America, with the last Giant Ground Sloths dying out. All Equidae become extinct in North America
10 ka	The Holocene Epoch starts 10,000 years

	ago after the Late Glacial Maximum. The last mainland species of Woolly mammoth (*Mammuthus primigenius*) die out, as does the last *Smilodon* species
6 ka	Small populations of American Mastodon die off in places like Utah and Michigan
4500 ya	The last members of a dwarf race of Woolly Mammoths vanish from Wrangel Island near Alaska
384 ya (1627)	The last recorded wild Aurochs die out
75 ya (1936)	The Thylacine goes extinct in a Tasmanian zoo, the last member of the family Thylacinidae

WE ARE UNIQUE AS A RESULT OF OUR HISTORY AND EVOLUTION

Life in the Universe and Where to find it

Bibliography

Anderson, P., *Is There Life on Other Worlds?*, Collier Books, (1963)

Brown, R.H., Lovell, A.C.B., *The Exploration of Space by Radio,* Wiley, (1958)

Cameron, A.G.W., *Interstellar Communications,* Benjamin, (1963)

Christiansen, W.N.,Hogbom J.A., *Radiotelescopes,* Cambridge, (1969)

Dessler, A.J. (Ed.), *Physics of the Jovian Magnetosphere,* Cambridge, (1983)

Feinberg, G.,Shapiro, R., *Life Beyond Earth,* Morrow, (1980)

Ferris, T., *The Mind's Sky,* Bantam, (1992)

Grinspoon, D., *Lonely Planets,* Harper-Collins, (2003)

Kirk, G.S., Raven, J.E., *The Presocratic Philosophers*, Cambridge, (1966)

Kraus, J.D. *Radio Astronomy,* Cygnus-Quasar, (1986)

Lemonick, M.D., *Other Worlds,* Touchstone, (1999)

MacGowan, R.A., Ordway III, F.I., *Intelligence in the Universe*, Prentice-Hall, (1966)

Munitz, M.K., *Theories of the Universe,* Free Press, (1957)

Penrose,R.,Hameroff,S.,Kak,S., *Consciousness in the Universe,* Cosmology Science Publishers, (2011)

Rowan-Robinson, M., *Cosmic Landscape,* Oxford University Press, (1979)

Shklovskii, I.S., Sagan, C., *Intelligent Life in the Universe,* Delta, (1966)

Steinberg, J.L., Lequeux, J., *Radio Astronomy,* McGraw-Hill, (1963)

Thompson, A.R., Moran, J.M., Swenson, G.W., *Interferometer and Synthesis in Radio Astronomy,* Wiley-Interscience, (1986)

Unsold, A., *The New Cosmos*, Springer-Verlag, (1969)

Verschuur, G.L., *The Invisible Universe,* Heidelberg, (1974)

Zubrin, R., *Entering Space,* Archer-Putnam, (1999)

References

1. Wikipedia, from June 2011 to October 2011, various articles
2. Project Cyclops Report
3. Aerofiles.com
4. www.daviddarling.info/encyclopedia/H/HRMS.html
5. journalofcosmology.com/Commentary207.html

Wikipedia sources used in the text:

en.wikipedia.org/wiki/Magnetosphere_of_Jupiter

en.wikipedia.org/wiki/Extrasolar_planet

en.wikipedia.org/wiki/Abiogenesis

en.wikipedia.org/wiki/Timeline_of_evolution

http://en.wikipedia.org/wiki/List_of_timelines

http://en.wikipedia.org/wiki/Cosmic Pluralism

Other Internet Sources:

http://journalofcosmology.com/

http://journalofcosmology.com/Commentary207.html

http://vaedrah.angelfire.com/seti_rx.htm

http://www.satsig.net/seticalc.htm

http://www.daviddarling.info/encyclopedia/H/HRMS.html

http://www.daviddarling.info/encyclopedia/T/target_star.html

http://newscenter.berkeley.edu/2011/05/13/uc-berkeley-seti-survey-focuses-on-kepler's-top-earth-like-planets/

http://seti.berkeley.edu/seti_at_the_gbt

http://www.nature.com/news/2009/091001/full/news.2009.966.html - Ardi, oldest huminoid, 4.4

http://www.used-robots.com/robot-education.php?page=robot+timeline

http://wiki.answers.com/Q/What_is_the_lowest_temperature_any_livin g_thing_can_survive_at#ixzz1PJiqZr44

http://www.sciencedaily.com/releases/2005/08/050809064541.htm

http://www.atlasoftheuniverse.com/50lys.html

http://www.phys.unm.edu/%7Elwa/lwatv/55865.mov

Other Sources:

Drake, F.D. "Project Ozma," *Physics Today*, 14,140 (1961).

Drake, Frank, "Project Ozma: The Search for Extraterrestrial Intelligence," Proceedings of the NRAO Workshop held at the National Radio Astronomy Observatory, Green Bank, West Virginia, Workshop No. 11, May 20-22, Kellerman, K.I. And Selelstad, G.A. Eds., p.23 (1985)

Gurnett, D. A., *Planetary Radio Emissions,* Astronomy and Astrophysics Encyclopedia,(1992)

Project Cyclops, NASA study CR 114445, 1971

ABOUT THE AUTHOR

Kevin Shoemaker was born in New York City in April of 1954. A son of an actress and musician turned professor. He has lived in several states and has been educated in the fields of philosophy, radio astronomy and antenna design. He has authored several technical papers in astronomy and has nine patents in the fields of aviation, antenna design and meteorology. In addition, he is an avid pilot and boat owner and holds several certificates for operating airplanes, helicopters and performing flight instruction. Currently he works as an antenna and radar designer at Cape Canaveral. Mr. Shoemaker is a father of one daughter and one son and lives near Boulder with his wife, Judi.

Comments?   e-mail: Shoemakerlabs@gmail.com

Other books by the author:

Mars Life
Practical Antenna Design
The Voyages of Gaea
Sunrise Descending

www.ingramcontent.com/pod-product-compliance
Lightning Source LLC
Chambersburg PA
CBHW022050210326
41519CB00054B/296